PRAISE FOR
Ending Parkinson's Disease

"This book shines an essential spotlight on the need to end Parkinson's, illuminating the key issues we are all facing. It is the role of everyone in the Parkinson's community, patients, researchers, clinicians and advocates, to share this book and ensure that the authors' voices are heard."

—HELEN MATTHEWS, Deputy CEO,
The Cure Parkinson's Trust

"Just like music is about the people, *Ending Parkinson's Disease* is about the scores of patients and families, like my mother-in-law, who battled each day against the many challenges of neurodegenerative diseases. We need to synchronize the worldwide orchestra of those affected in order to bring an to end to these diseases. This book will provide the roadmap."

—CHRISTIAN MCBRIDE, multiple
Grammy Award–winning bassist

"A provocative read. It will certainly encourage you to consider what more you can be doing in the fight against Parkinson's Disease."

—LESLIE CHAMBERS, MSPH, President and CEO
of American Parkinson Disease Association, APDA

"Parkinson's disease is not one disease but rather a collection of many with different contributing factors, and it satisfies many of the criteria of a pandemic. While the authors of *Ending Parkinson's Disease* are hopeful about making patients' lives better, their real passion is clearly centered around preventing people from ever having to face the disease in the first place. The book is a real eye-opener for people with Parkinson's, caretakers, and healthcare professionals, and should be used by activists all over the world in their discussions with politicians, policy makers and budget holders."

—SUSANNA LINDVALL, Vice President,
European Parkinson's Disease Association, EPDA

"As someone with a family history of Parkinson's disease, I have been looking for a bold and actionable statement to help the millions affected. This book is it."

—JUSTIN MCARTHUR, MBBS, MPH, Director,
Department of Neurology, Johns Hopkins Medicine

ENDING
PARKINSON'S
DISEASE

ENDING

PARKINSON'S

DISEASE

A PRESCRIPTION
for ACTION

Ray Dorsey, MD ✦ Todd Sherer, PhD
Michael S. Okun, MD ✦ Bastiaan R. Bloem, MD, PhD

PUBLICAFFAIRS
NEW YORK

PublicAffairs
Hachette Book Group
1290 Avenue of the Americas, New York, NY 10104
www.publicaffairsbooks.com
@Public_Affairs

Printed in the United States of America
First Trade Paperback Edition: 2021

Published by PublicAffairs, an imprint of Perseus Books, LLC, a subsidiary
of Hachette Book Group, Inc. The PublicAffairs name and logo is a trademark
of the Hachette Book Group.

The Hachette Speakers Bureau provides a wide range of authors for speaking
events. To find out more, go to www.hachettespeakersbureau.com or call
(866) 376-6591.

The publisher is not responsible for websites (or their content) that are not
owned by the publisher.

Print book interior design by Linda Mark

Illustrations and A Prescription for Action section design by
Gerardo Torres Davila and Ellen Wagner

Library of Congress Cataloging-in-Publication Data
Names: Dorsey, Ray, author.
Title: Ending Parkinson's disease : a prescription for action / Ray Dorsey,
    Todd Sherer, Michael S. Okun, Bastiaan R. Bloem.
Description: New York : PublicAffairs, [2020] | Includes bibliographical
    references and index.
Identifiers: LCCN 2019030265 | ISBN 9781541724525 (hardcover) |
    ISBN 9781541724495 (ebook)
Subjects: LCSH: Parkinson's disease.
Classification: LCC RC382 .D644 2020 | DDC 616.8/33—dc23
LC record available at https://lccn.loc.gov/2019030265

ISBNs: 978-1-5417-2452-5 (hardcover), 978-1-5417-2449-5 (ebook),
    978-1-5417-2450-1 (paperback)

LSC-C

Printing 4, 2023

*To those who bear the burden of Parkinson's disease
and to those who will help end it*

Contents

Note to Readers

IN THIS BOOK, WE SHARE THE STORIES OF PEOPLE AFFECTED BY Parkinson's disease. Most of these accounts are based on interviews we conducted. In some cases, individuals asked that their names be changed to protect their privacy; we note these instances in the text. Other stories are taken from published reports and are referenced as such.

The views expressed here are those of the authors and not necessarily those of their employers. The authors are devoting their net proceeds to efforts to help end Parkinson's.

Glossary

Below are definitions for key terms used in this book.

alpha-synuclein: A protein that is misfolded, or altered, in people who have Parkinson's disease. The misfolded protein forms clumps in nerve cells and likely contributes to nerve cell death.

dopamine: A chemical that is released from nerve cells in areas of the brain that are affected by Parkinson's.

levodopa: A drug that is converted into dopamine and is a highly effective medication for Parkinson's disease.

Lewy bodies: The clumps of misfolded alpha-synuclein and other proteins that are found in the brains of individuals with Parkinson's disease.

LRRK2: A specific gene that codes, or gives building instructions, for a protein in the brain and other parts of the body. Mutations in this gene are the most common genetic cause of Parkinson's disease.

mitochondria: The energy-producing parts of cells, which are damaged in Parkinson's and by some pesticides.

MPTP: An accidental chemical by-product of a street formulation of synthetic heroin. MPTP kills dopamine-producing nerve cells and has caused parkinsonism in some users of heroin.

neurotransmitter: A chemical that is released from nerve cell endings and enables communication between cells.

parkinsonism: A general term for any syndrome that causes tremors, slowed movements, stiffness, and imbalance. This condition has many causes, including Parkinson's disease, certain medications, and other diseases.

pesticide: Any substance used to prevent, destroy, repel, or mitigate any pest, including herbicides (for weeds), insecticides (for insects), and fungicides (for fungi) among other chemicals.

substantia nigra: A Latin phrase that literally means "black substance." It refers to a small region of the brain that contains pigmented dopamine-producing nerve cells, which are damaged in people who have Parkinson's disease.

Abbreviations

Below are common abbreviations used in this book.

EPA: US Environmental Protection Agency

FDA: US Food and Drug Administration

NIH: US National Institutes of Health

TCE: Trichloroethylene, a chemical that has been linked to Parkinson's disease

ENDING
PARKINSON'S
DISEASE

INTRODUCTION

> Every civilization has its own kind of pestilence and can control it
> only by reforming itself.
>
> —René Dubos, *Mirage of Health*, 1959[1]

ON A BRILLIANT, BLUE-SKY DAY IN JUNE 2018, THE UNIVERSITY
of Rochester hosted its annual Men's Health Day at the Locust Hill
Country Club in upstate New York. Over three hundred men, most in
their fifties, sixties, and seventies, came to hear the latest on enlarged
prostates, colon cancer, and heart disease. I came to speak about Parkinson's disease.

Months earlier I had written a paper titled "The Parkinson Pandemic" with my friend and colleague—and now coauthor—Bas
Bloem.[2] In it, we explained that neurological disorders are the world's
leading cause of disability. And the fastest growing of these conditions is not Alzheimer's but Parkinson's disease. From 1990 to 2015,
the number of people with Parkinson's more than doubled from 2.6
million to 6.3 million.[3] By 2040, the number will double again to at
least 12.9 million, a stunning rise (**Figure 1**).[4]

This is what I know. This is what I study. But as I stood there in
front of the packed room at Men's Health Day, I was not prepared for
what I was about to see. I opened my talk by asking how many people

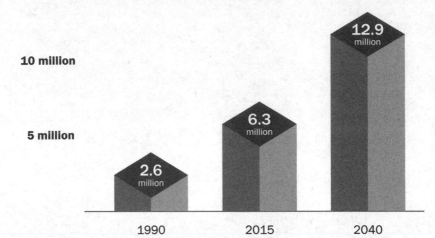

FIGURE 1. Estimated and projected number of people with Parkinson's disease globally, 1990–2040.[5]

in the audience had a friend or family member with Parkinson's. Before I could finish asking the question, over two hundred hands had flown up—almost the entire room. Everyone looked around. A silence settled on us as we took in the sight. It didn't matter that I was an expert or that I had helped develop the statistics. Data always feel remote, but here it was in front of me, the evidence of the pandemic.

Parkinson's disease is characterized by tremors, slowness in movement, stiffness, and difficulties with balance and walking. It can also cause a wide range of symptoms that are not visible—loss of smell, constipation, sleep disorders, and depression. Most people with Parkinson's are diagnosed in their fifties or later. But it is not just a disease of the elderly. Up to 10% of those with the condition develop the disease in their forties or younger.

Parkinson's stems from a loss of nerve cells in a particular region of the brain that produces dopamine, the brain chemical that helps control movements such as walking. The disease has multiple causes including environmental hazards—air pollution, some industrial solvents, and particular pesticides. In addition, certain genetic mutations, head trauma, and the lack of regular exercise all increase risk.[6]

The scale of the disease can feel overwhelming and the challenge daunting. But we can stop Parkinson's in some cases, and we may already know how.

In the meantime, while there is no cure for Parkinson's yet, many aspects of it are treatable. Just as exercise can reduce our risk of developing the disease, it can also help alleviate its symptoms.[7] Medications aimed at replacing the dopamine that is lost in the brain are also beneficial. However, complications can develop with high doses or long-term use of some drugs. In certain cases, brain surgery can help treat these side effects.[8]

Although Parkinson's is a progressive disorder—it becomes worse over time—most people can still live long and productive lives. Especially for the first five to ten years following diagnosis, individuals can function at high levels, working, traveling, and enjoying life.

Of course, the disease still takes a profound toll on individuals and their families. Up to 40% of people with Parkinson's will eventually require nursing home care, and the caregiving burden is immense.[9] Life expectancy is reduced modestly, and many die from falls or pneumonia.[10]

☿ ☿ ☿

The seminal description of Parkinson's disease came in 1817, at the height of the Industrial Revolution in London.[11] Dr. James Parkinson observed six individuals who walked with an unusual gait and had "shaking limbs." Parkinson's disease, as it became known, was almost certainly rare then.

Neither our increased awareness of the disease nor our lengthening lifespans can fully account for the upsurge in diagnoses that we now face. Our knowledge of another neurological disorder, multiple sclerosis, has increased too, and we have improved diagnostic tools for it. Rates for multiple sclerosis have indeed gone up, but that increase is nothing like the exponential rise of Parkinson's (**Figure 2**). As for aging, more people are, of course, living longer. For example, from 1900 to 2014, the number of individuals over age sixty-five in the United

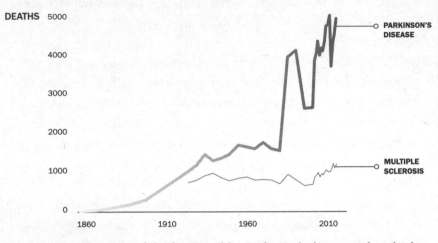

FIGURE 2. Number of deaths caused by Parkinson's disease and multiple sclerosis in England, 1860–2014.[12] Changes in coding in the 1980s likely contributed to the fluctuations in deaths recorded during this period.

Kingdom increased about sixfold. However, over that same period, the number of deaths due to Parkinson's disease increased almost three times faster.

How did we get here? While industrialization has raised incomes and life expectancies around the world, its products and by-products are also likely increasing the rates of Parkinson's.[13] Air pollution began to worsen in England in the eighteenth century, metal production and its harmful fumes increased in the 1800s, the use of industrial chemicals rose in the 1920s, and synthetic pesticides—many of which are nerve toxins—were introduced in the 1940s.[14] All are linked to Parkinson's—people with the most exposure have higher rates of the disease than the general population.

The evidence for this connection is overwhelming. Countries that have experienced the least industrialization have the lowest rates of the disease, while those that are undergoing the most rapid transformation, such as China, have the highest rates of increase.[15] Specific metals, pesticides, and other chemicals have all been tied to Parkinson's in numerous human studies.[16] When animals are exposed to many of these substances in lab experiments, they develop the typical characteristics of the disease, including difficulty walking and tremors.[17]

Despite the vast evidence, we are doing little to manage these threats. The US Environmental Protection Agency (EPA) had at one time proposed banning one of the chemicals that is tied to Parkinson's, a solvent called trichloroethylene. But after lobbying by the chemical industry, the EPA decided in 2017 to postpone the ban indefinitely.[18] The uses of trichloroethylene have been so numerous and widespread—in washing away grease, cleaning silicon wafers, removing spots in dry cleaning, and even, until the 1970s, decaffeinating coffee—that almost all of us have been exposed to it at some point in our lives.[19] Some of these uses continue today. Almost half of Superfund sites—land so polluted that the EPA or the responsible parties have to clean it up—which are found in nearly every state, are contaminated with trichloroethylene.[20] Thousands of other sites are polluted across the country, including one, as I discovered in the process of writing this book, fifteen minutes from my home.[21]

As a result, up to 30% of the US drinking water supply has been contaminated with trichloroethylene.[22] Because it readily evaporates from groundwater and soil, the solvent, like radon, can enter homes or offices through the air, undetected.[23] Parkinson's is not even the most concerning safety risk. According to the EPA, trichloroethylene also causes cancer.[24]

But trichloroethylene is only one dangerous chemical that we have failed to protect ourselves against. Paraquat is a pesticide that is so toxic that thirty-two countries, including China, have banned it.[25] Exposure to the chemical increases the risk of Parkinson's by 150%.[26] Yet the EPA has done little. And as the agency charged with protecting our environment sits, paraquat's use on US agricultural fields has doubled over the last decade.[27]

The nerve toxin chlorpyrifos is the most widely used insecticide in the country, drenching golf courses and dozens of crops, including almonds, cotton, grapes, oranges, and Washington State apples. It has been linked not only to Parkinson's but also to problems with brain development in children. Again, the EPA has shelved a ban. When a federal court stepped in to take action against the chemical, the Trump administration appealed.[28] And in July 2019 in response to a court ordering a final ruling, the EPA decided that it would allow continued use of chlorpyrifos.[29]

All of the evidence indicates that the full effect of the Parkinson's pandemic is not inevitable but, to a large extent, preventable. However, we cannot remain silent.

We have been here before. We have faced down other difficult illnesses that have threatened us. Three of them—polio, HIV, and breast cancer—share similarities with Parkinson's and offer valuable lessons for how we can take it on. Polio is a disabling neurological condition. HIV affected large numbers of individuals globally in a very short period. Breast cancer likely has both environmental and genetic causes.[30] At some point, society ignored all three until the people who knew the diseases intimately—and understood their toll firsthand—stepped forward. Their activism changed the courses of these diseases and has improved and saved the lives of millions.

That is why we are writing this book. Yes, we are sounding the alarm that this pandemic is upon us. But we also know that if we respond now to the challenge it presents, we can save many people from suffering. Individually and collectively, we can take some very practical actions to stop the damage.

In *Ending Parkinson's Disease*, we will discuss what new policies, protections, and resources can slow the disease. The Netherlands, for example, banned trichloroethylene, paraquat, and other pesticides linked to Parkinson's years ago—and it worked. Rates of the disease decreased.[31] This outcome shows how stemming the tide of Parkinson's is within our reach.

We will also examine how we can offer better support and care to the millions affected by Parkinson's today. We will see what new therapies are on the horizon and how close we are to introducing novel treatments that may slow or stop the progression of Parkinson's. Some of these will arrive in time to help people who already have the disease. Others may even help to prevent Parkinson's altogether.

At the end of the book, we outline what all of us can do to lower our risk, increase resources to address the condition, extend expert care to all those in need, and slow the advance of Parkinson's.

Along the way, we will highlight the experiences of courageous individuals with the disease, tireless caregivers, and fearless advocates. We will hear their stories, learn from their experiences, and take inspiration from their actions.

The four of us—one neuroscientist and three neurologists who specialize in Parkinson's—have devoted most of our professional lives to this disease. Twenty years ago, Dr. Todd Sherer conducted groundbreaking research linking pesticides to Parkinson's disease. He now leads The Michael J. Fox Foundation for Parkinson's Research, the largest private funder of Parkinson's research in the world.[32] Dr. Michael Okun, who first characterized Parkinson's as a pandemic, has pioneered new surgical treatments for people with the disease and written multiple books and articles on the topic.[33] Professor Bas Bloem is a leading authority on gait disturbances and falls in Parkinson's and co-created the world's largest care program for people with the disease.[34] And with my colleagues, I have used technology to expand access to care and develop new methods for measuring the disease.[35] All of us are working to advance better treatments for the condition.

While we are hopeful about making our patients' lives better, our true passion is preventing people from ever having to face Parkinson's. We are frustrated when we see women and men in our clinics who have suffered head trauma or been exposed to pesticides on a farm, solvents at work, contaminated groundwater in their neighborhoods, or polluted air in their homes. All of these risks for Parkinson's can be mitigated. We humans have helped create this pestilence. And we can now work to end it.

PART ONE

A Formidable Disease

SIX MEN IN LONDON

The Discovery of a New Disease and Its Causes

The unhappy sufferer has considered [the disease] as an evil,
from the domination of which he had no prospect of escape.

—Dr. James Parkinson, "An Essay on the Shaking Palsy," 1817[1]

AT THE TURN OF THE NINETEENTH CENTURY, BRITAIN WAS booming and its Industrial Revolution was about to transform the world. Coal mining fueled James Watt's steam engine. Iron smelting enabled the building of new bridges, while steamships and the telegraph linked disparate lands. The spinning jenny churned out wool and cotton, gas lights illuminated theaters, and author Jane Austen challenged social norms. Populations soared, and London, the epicenter of all of it, exploded with prosperity.[2]

The city was also becoming filthy. Humans and factories dumped their waste into the River Thames. Poor sanitation and overcrowding spread infectious diseases, including cholera, typhus, and tuberculosis. With the new industries came new chemicals and pollutants spewing from what the poet William Blake called "dark Satanic mills."[3]

According to one environmental researcher, "It's difficult to fully capture just how polluted London's air was throughout the 19th century."[4] These industrial London fogs (**Figure 1**) were "often so dense that they . . . interrupted general economic activities, and even contributed

FIGURE 1.   Illustration of London fog, 1847.

to [the city's] becoming a breeding ground for crime."[5] It was on these hazy streets that a seasoned physician observed something new.

A BRIEF HISTORY OF PARKINSON'S DISEASE

A suffragist, activist, paleontologist, and advocate for the mentally ill, Dr. James Parkinson had many lives.[6] Because of his politically radical stances, he used pseudonyms and, at one point, narrowly avoided imprisonment for his alleged role in a plot to assassinate King George III.[7] However, his most enduring contribution to humanity was not his politics but a single essay that was destined to become a medical classic.

In 1817, Parkinson was a local doctor in Hoxton Square, London, where, almost two hundred years earlier, William Shakespeare had crafted many of his plays. Parkinson's literary contribution was titled "An Essay on the Shaking Palsy." By this time, he already had a wealth of clinical experience, gathered over more than thirty-two years of patient care.[8] In his essay, Parkinson described six men, three of whom he had simply noticed on the street, who all shared similar characteristics—tremors, a bent posture, an abnormal walk, and a tendency to fall.[9]

Although ancient Chinese, Egyptian, Greek, and Indian texts provide rare depictions of some of these same symptoms, Parkinson's essay was the most substantive.[10] As he indicated, tremors had long been known and had multiple causes. However, the multi-symptom affliction that Parkinson was now observing on his walks had yet to be classified.[11] His essay was well received, but its importance would not be recognized for decades.[12]

Fifty years after James Parkinson's essay (**Figure 2**), Dr. Jean-Martin Charcot, the famous French neurologist, called the disorder "la maladie de Parkinson," or Parkinson's disease.[13] Charcot added slow movements and stiffness to the list of key features.[14] He also noted that not everyone with Parkinson's disease had a tremor.

By the end of the nineteenth century, the clinical features of Parkinson's disease were well known. In his 1892 medical textbook, Sir William Osler, the father of modern medicine, wrote, "When well established, the disease is very characteristic, and the diagnosis can be made at a glance."[15] While the external features of Parkinson's were obvious, the underlying biological alterations were not.

THE ROLE OF DOPAMINE

What Parkinson and Charcot could not have observed in their time were the changes occurring in the brains of their patients. Scientists had long overlooked, downplayed, and ignored the chemical dopamine. But Dr. Arvid Carlsson, a Swedish pharmacologist working in the 1950s, saw dopamine differently.[16] In experiments, he established that dopamine allowed nerve cells to communicate with one another. In other words, it was a neurotransmitter.

Carlsson also showed that a brain region important for movement contained high levels of dopamine. To demonstrate the chemical's importance, he gave rabbits a drug that lowered dopamine levels in the brain. The rabbits lost their ability to hop and simply lay down. When

1817
England
Dr. James Parkinson
writes
seminal essay on the
"Shaking Palsy"

1912
Germany
Dr. Fritz Lewy discovers
clumps of misfolded protein
in brain regions affected by
Parkinson's disease

1800

1900

1872
France
Dr. Jean-Martin Charcot
details the condition and
renames it
"Parkinson's disease"

1886
England
Sir Williams Gowers
produces a two-volume
"bible of neurology" that
describes Parkinson's
disease

FIGURE 2. Two hundred years of Parkinson's disease, 1817–2017.

15

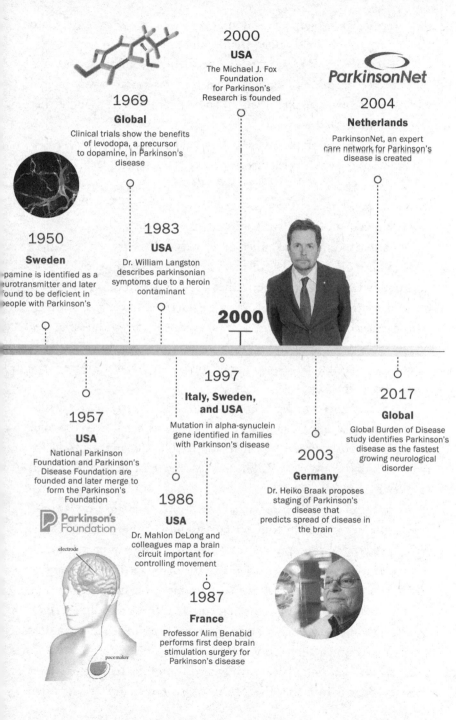

1950

Sweden

pamine is identified as a eurotransmitter and later found to be deficient in people with Parkinson's

1969

Global

Clinical trials show the benefits of levodopa, a precursor to dopamine, in Parkinson's disease

1983

USA

Dr. William Langston describes parkinsonian symptoms due to a heroin contaminant

2000

USA

The Michael J. Fox Foundation for Parkinson's Research is founded

2000

ParkinsonNet

2004

Netherlands

ParkinsonNet, an expert care network for Parkinson's disease is created

1957

USA

National Parkinson Foundation and Parkinson's Disease Foundation are founded and later merge to form the Parkinson's Foundation

Parkinson's Foundation

1986

USA

Dr. Mahlon DeLong and colleagues map a brain circuit important for controlling movement

1987

France

Professor Alim Benabid performs first deep brain stimulation surgery for Parkinson's disease

1997

Italy, Sweden, and USA

Mutation in alpha-synuclein gene identified in families with Parkinson's disease

2003

Germany

Dr. Heiko Braak proposes staging of Parkinson's disease that predicts spread of disease in the brain

2017

Global

Global Burden of Disease study identifies Parkinson's disease as the fastest growing neurological disorder

electrode

pacemaker

they were given levodopa, a drug that is converted by nerve cells into dopamine, the rabbits regained their bounce.[17]

Carlsson presented his research to the scientific community in 1960. He expected his peers to be excited. Instead, his research was greeted with almost universal skepticism.[18] Some doctors even thought that levodopa might be a poison.[19]

Although stung at first, Carlsson would later say, "I am pleased [when] people are saying they don't believe in me. Then I feel I'm probably on the right track."[20] And he was. His efforts and perseverance are the basis for what is still the most effective treatment for Parkinson's and resulted in a Nobel Prize in 2000.

Notwithstanding the skeptics, other researchers picked up where Carlsson left off.[21] They began measuring dopamine in the brains of deceased individuals. They found that levels of dopamine in the brains of people who had had Parkinson's were ten times lower than in those without the condition. The levels were especially low in the same area of the brain that Carlsson had identified as typically dopamine rich. There was a simple relationship: the lower the dopamine levels, the worse the symptoms.[22]

This area of the brain that is normally dopamine rich is called the substantia nigra, Latin for "black substance." It gets its name from the color of a pigmented chemical found in the region's dopamine-producing nerve cells. In people with Parkinson's disease, these nerve cells die off (**Figure 3**).

As it turns out, Parkinson's affects more than just dopamine-producing nerve cells in the substantia nigra. Other regions of the brain producing different neurotransmitters also suffer cell loss.[23] This additional damage is responsible for many of the symptoms of Parkinson's that are not related to movement or "motor" function, such as disturbed sleep, anxiety, pain, and thinking difficulties.[24] Some of these symptoms can be even more disabling than the motor symptoms controlled by dopamine.[25]

Based on Carlsson's breakthrough with rabbits, researchers later tried levodopa in humans. The results were spectacular.[26] "Bed-ridden

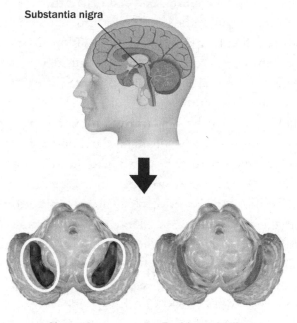

FIGURE 3. The substantia nigra (Latin for "black substance") in individuals without and with Parkinson's disease. In Parkinson's, the substantia nigra is one of the key regions in the brain where cells die.

patients who were unable to sit up, patients who could not stand up when seated, and patients who when standing could not start walking performed all these activities with ease after [levodopa]," the doctors wrote. "They walked . . . and they could even run and jump."[27] Multiple clinical trials would later replicate these dramatic effects.[28] Dr. George Cotzias, a Greek American scientist who led many studies of levodopa, called it a "true miracle drug . . . of our age."[29]

A DETECTIVE STORY

A dopamine deficiency was now understood to cause many of the symptoms of Parkinson's. But no one knew what killed the nerve cells that produced it. Dr. Parkinson had speculated that the disease was caused by compression of the lowest part of the brain.[30] Sir William Osler, one of the founding professors of Johns Hopkins Hospital, theorized that "exposure to cold and wet and business worries and anxieties" were possible causes.[31] Neither was right. The first real insights came from an unlikely source.

On July 16, 1982, at a hospital in San Jose, California, a neurology resident interrupted Dr. William Langston's morning cup of coffee. "Dr. Langston, you have to come down here," he said. "I've never seen anything like it, and no one is sure what this patient has."[32]

George Carillo, a forty-two-year-old man with a history of substance abuse, had just been admitted to a locked psychiatric unit. According to Langston, "The patient's condition was indeed extraordinary. He was clearly awake, but had virtually no spontaneous movement. . . . [He] looked like a textbook case of advanced [Parkinson's disease] before the days of levodopa. But this case didn't fit. . . . He was in his early forties, and his symptoms came on literally overnight. We had a first class 'medical mystery' on our hands."[33]

Spurred by this mystery, Langston and his colleagues checked news reports and reached out to police to see if there were any other individuals like Carillo. They soon learned of five people with similar symptoms who also lived in Northern California. They had no apparent connection until Langston, digging into their histories, realized that all of them had used a new synthetic heroin that had recently hit the streets in several Northern California cities.[34] The six users had developed "virtually all of the motor features of typical Parkinson's disease, including [a] tremor."[35]

Unlike the usual presentation of Parkinson's disease, which has a gradual onset, the first symptoms in these individuals came on rapidly and were very severe. Within two weeks, they had developed "near total immobility . . . a complete inability to speak intelligibly, a fixed stare . . . [and] constant drooling."[36] It normally takes people with Parkinson's years to reach such an advanced stage of the disease.

Levodopa helped all the patients. However, they all required life-time treatment, soon developed marked complications, and lived with considerable disability. Langston described their fate: "Try to imagine never being able to move normally again. Never being able to raise your arm when you wanted to. Never being able to walk to the dinner table. Your life permanently and forever changed."[37] Connie Sainz was in that original California group and was only twenty-five when she developed her symptoms. She lived with the condition for thirty-six years before her death in 2018.[38]

Langston wanted to know what caused their disability. After obtaining samples of the synthetic heroin from police raids and friendly dealers, Langston and his colleagues found that the drug contained a chemical called MPTP.

The creation of MPTP was not intentional. Its makers were trying to synthesize an opioid similar to the prescription drug Demerol to sell on the streets. In the process, they accidentally produced MPTP.[39] In the brain, it is converted to a compound called MPP+, which kills nerve cells in the substantia nigra—the cells that die off in Parkinson's. Langston and his colleagues had found the first cause of parkinsonism (**Box A**).

BOX A. WHAT IS PARKINSONISM?

Parkinsonism is any syndrome that results in a combination of tremors, slow movements, rigidity, and imbalance. This collection of symptoms has many causes, including MPTP, some antipsychotic medications, infections, and neurological diseases. Parkinson's disease is the most common cause of parkinsonism. Other diseases can also bring about these symptoms, but they usually come with additional features of their own.

The name "Parkinson's disease" is generally restricted to parkinsonism that (a) contains few features of other diseases that can cause the syndrome and (b) responds well to levodopa.[40]

As it turns out, Langston's patients in California were not the first to develop a Parkinson's-like syndrome in response to MPTP.[41] Another victim, Barry Kidston, was identified several years earlier, but despite its tremendous importance, his story was almost lost in an obscure journal.[42] In 1966, Barry was fourteen years old when he fractured both wrists and a leg in a car accident. For pain relief, the doctors prescribed the opioid Demerol. Barry quickly became dependent and soon started abusing many drugs.[43]

Ten years after the accident, he was a chemistry major at George Washington University. According to his mother, Geraldine, he was outgoing, had a great personality, and liked to help people. Unfortunately, he was also still addicted to drugs.

He wanted to find a way to get sober. Rather than go to rehab, he started a laboratory in his parents' basement.[44] The idea, he told his mother, was to develop a nonaddictive drug—something along the lines of methadone, which would later become more widely available—that would help him kick the habit that had followed him all through school.

Barry checked out chemistry books from the nearby National Institutes of Health (NIH) and the Bethesda Naval Hospital libraries. The science was tricky and did not go well. He continued experimenting until one day he lifted his arms over his head, maybe to stretch, and could not bring them back down.[45] According to his mother, "He had to walk round with them stuck straight up."[46]

An astute neurologist realized that Barry's symptoms, which also included a marked slowing of movement and a tremor, added up to parkinsonism.[47] The neurologist treated him with levodopa, which quickly unfroze him. Because the case was unusual, Barry was referred to the NIH for another opinion.

There, he was seen by a team of physicians and scientists, including Dr. Eric Caine, now a professor of psychiatry at the University of Rochester. Caine recalled seeing Barry as a young man who was "stiff as a board" and "extremely slow." The situation, he said, was very sad. "You have this guy savvy enough to do sophisticated chemistry who now has a terrible prognosis." Upon questioning, the cause of his parkinsonism seemed to be related to the experiments that Barry was conducting and the drugs that he was producing—and consuming.

An NIH chemist went to the Kidstons' home to retrieve some of Barry's laboratory equipment. The researchers wanted to replicate his basement experiments in the lab. They were able to produce chemicals that were identical to the traces of material they found on Barry's laboratory glassware.[48] The scientists concluded that his symptoms were probably due to a "sloppy batch" of the opioids that he was synthesizing.[49] Eventually, the culprit of Barry's symptoms was identified as MPTP, the same ingredient that would later sicken the group in California.[50]

Unfortunately, Barry did not stop abusing drugs. On a September morning in 1978, he called a friend to pick him up from his parents' home. He left, and it was the last time they saw him. "He walked out the door through the car port, smiling, and waving goodbye," his mother said.[51]

At about 10 p.m., the doorbell rang. Outside were three gentlemen: a state trooper, an FBI agent, and a coroner. Barry Kidston had been found dead from a cocaine overdose under a tree on the NIH campus.[52] An autopsy of his brain later showed destruction of nerve cells in the substantia nigra.[53]

In 1985, the Kidstons shared their story in testimony to Congress.[54] They wanted other parents to be spared the loss they experienced and hoped that something good would come from their son's tragedy.

Something did. According to Langston, out of the Kidstons' devastation, "a renaissance in the epidemiology of [Parkinson's] disease" took place.[55] Scientists began searching for other environmental causes.

They soon realized that the chemical structure of the toxic MPP+ (produced from MPTP) was remarkably similar to a pesticide called paraquat.[56] This structural similarity led Langston in California and others to find additional chemicals linked to Parkinson's. These newly identified chemicals included the pesticides rotenone and

Agent Orange, which eviscerated the jungles of Vietnam during the war in that country.[57]

The discovery of MPTP's role in parkinsonism had another benefit. Langston realized that it would allow researchers to create an animal model of the disease (**Box B**), which did not previously exist.[58] Quite remarkably, no animal in nature spontaneously develops Parkinson's disease. In Langston's words, it is "a uniquely human disease."[59] But now MPTP could cause the symptoms in animals and enable the testing of new treatments on them. Langston and colleagues thus used the harmful effects of MPTP to push Parkinson's research forward.[60]

In the 1990s, researchers at Emory University were investigating whether the common pesticide rotenone could cause Parkinson's. Rote-

---

BOX B. WHAT IS AN ANIMAL MODEL?

Scientists use animals to study diseases, including Parkinson's. In some cases, diseases occur naturally in animals.[61] For example, dogs can spontaneously develop high blood pressure, heart failure, or diabetes. Other diseases, such as Parkinson's, are unique to humans. So in order to study these illnesses, scientists give them to the animals, often mice or rats. This can be done chemically—as with MPTP—or by changing the animal's genes.

While animal models have contributed to many scientific advances, including vaccines for measles and insulin for diabetes, they have their limitations.[62] Despite their genetic similarities (mice and humans share 95% of the same genes), not all findings in mice translate to humans.

These experiments also come at a cost—the welfare and lives of animals. Three principles govern animal research. The first is that animals must not be used when approaches that do not require them are available and reliable. Second, the number of animals used should be minimized. Third, harm to animals should be mitigated.[63]

none is derived from plants and used to be sold as a household insecticide. Today it is still used by fisheries to eliminate invasive species.[64]

Dr. Timothy Greenamyre and two young neuroscientists, Drs. Ranjita Betarbet and Todd Sherer, gave the chemical to rats.[65] The animals then developed features of Parkinson's, including slow movements, an unsteady walk, a hunched posture, and the "shaking of one or more paws that was reminiscent of [a] rest tremor," according to the researchers. And when they examined the rats' brains, they saw other signs of the disease, including loss of dopamine-producing nerve cells. They suspected that long-term exposure to low levels of certain noxious chemicals may eventually lead to Parkinson's.[66]

With the finding that pesticides can cause Parkinson's in mice and rats, Langston sought to extend the work to humans. To do so, he joined Dr. Caroline Tanner, a Parkinson's specialist with a doctorate in environmental health sciences who is now at the University of California, San Francisco. They wondered whether exposure to rotenone and other pesticides resulted in a higher risk of Parkinson's among individuals in close contact with pesticides—farmers.[67]

They found that farmers who used specific pesticides, including paraquat and rotenone, were more than twice as likely to develop Parkinson's as those who did not. In many cases, the exposure occurred fifteen years or more before diagnosis. The results suggested that chronic exposure to certain pesticides could indeed lead to Parkinson's years or decades later.

THE HUNT FOR GENETIC FACTORS

Fifteen years after identifying the first environmental cause of Parkinson's, researchers discovered the first genetic one.[68] In 1997, Dr. Mihael Polymeropoulos from the NIH and his colleagues identified a mutation in the alpha-synuclein gene (**Box C**) in a large Italian family and three Greek families with Parkinson's disease. In humans, this gene encodes, or gives instructions for, the building of something called the alpha-synuclein protein. This protein helps move neurotransmitters

---

BOX C. WHAT IS A GENE?

A gene is a piece of DNA that gives instructions for building a protein. Proteins are the workers of a cell. They perform various functions, including providing structure to the cell, transporting molecules, fighting infections, and carrying out chemical reactions.

Humans have approximately 20,000 genes. Like any message, a gene is made up of a normal sequence or series of letters or symbols. Mutations change that sequence. Some of these changes can alter the directions used to assemble proteins. A gene can have multiple mutations that can vary widely in significance— from having no effect to causing disease or even death.

---

within nerve cells. Proteins are generally produced in a folded state, just like a pile of freshly ironed sheets. However, the mutation identified in the Mediterranean families caused their alpha-synuclein proteins to change shape and consequently misfold—messing up the neat pile of linens and causing Parkinson's.[69]

Both environmental and genetic factors can trigger proteins to misfold (**Figure 4**). When this occurs, misfolded proteins can become toxic to nerve cells and cause disease. Rather than help transport neurotransmitters, as it normally does, the misfolded alpha-synuclein forms clumps in nerve cells. This misfolding may also spread to other nerve cells, eventually causing more cell death.[70] This leads to Parkinson's disease.

Eighty years before Polymeropoulos identified the alpha-synuclein genetic mutation, Dr. Fritz Jakob Heinrich Lewy, a Jewish neurologist who later fled Nazi Germany, observed the damage it could do. Using microscopic techniques that were cutting-edge at the time, he examined the brains of deceased individuals who had had Parkinson's.[71] In those brains, he was the first to observe clusters of the misfolded pro-

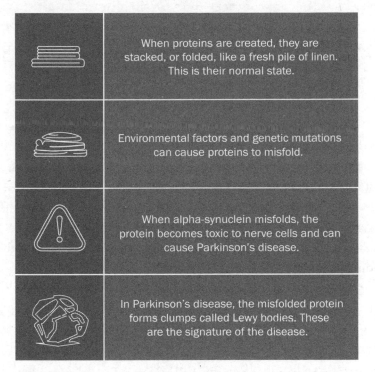

FIGURE 4. How a misfolded protein can lead to Parkinson's disease.

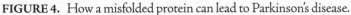

tein that was later identified as alpha-synuclein.[72] He noticed that the misfolded proteins formed clumps, which appeared to be the equivalent of garbage bags—all collected together ready for disposal—within nerve cells. These protein clusters became known as "Lewy bodies." They are found in the areas of the brain most affected by Parkinson's, including the substantia nigra. Present in almost all cases, Lewy bodies are now considered to be landmarks of the disease.[73]

While the alpha-synuclein genetic mutation is rare, the significance of its discovery was immense. It demonstrated that not all cases of Parkinson's are due only to the environment. In addition, it drove researchers to find numerous other genes that play a role in Parkinson's. Mutations in some of these genes, like those in alpha-synuclein, are sufficient to cause the disease. In other cases, mutations simply raise the risk of developing Parkinson's, a risk that often increases with age.

UNCOVERING A PERSONAL RISK

On September 18, 2008, Sergey Brin, the cofounder of Google, wrote a blog post titled "LRRK2."[74] Four years earlier, a research team had identified the most common genetic cause of Parkinson's disease— mutations in the LRRK2 gene.[75] Scientists had learned that 20% to 40% of Ashkenazi Jews and North African Arab-Berbers with Parkinson's disease had underlying LRRK2 mutations compared to 1% to 2% of the general population with Parkinson's.[76] For Brin, these findings were personal.

Brin's mother, who has the disease, "had always been haunted by Parkinson's because her aunt had suffered from it," Brin wrote. When his then wife, Anne Wojcicki, co-founded the direct-to-consumer genetic testing company 23andMe, Brin signed up as an early customer. He discovered that he and his mother carried the same LRRK2 mutation.[77]

Based on the genetic test results, Brin had a markedly higher risk— between 20% and 80%—of developing Parkinson's. He wrote,

> This leaves me in a rather unique position. I know early in my life something I am substantially predisposed to. I now have the opportunity to adjust my life to reduce those odds.... I also have the opportunity to perform and support research into this disease long before it may affect me.... I feel fortunate to be in this position. Until the fountain of youth is discovered, all of us will have some conditions in our old age only we don't know what they will be. I have a better guess than almost anyone else for what ills may be mine—and I have decades to prepare for it.[78]

In the years since, Brin and the foundation he started with Wojcicki have contributed well over $100 million to Parkinson's research.[79]

With environmental and genetic factors now identified as likely causes of Parkinson's, researchers sought to determine their relative importance. In 1999, Tanner and Langston in California conducted a pioneering study of more than 17,000 twin brothers—both fraternal and identical—from a World War II registry. They found that the likelihood that a twin brother had Parkinson's was similar among fraternal twins, whose genes are as similar as those of any two siblings, and identical twins, who have the same genes. The results suggested that environmental causes are more important in Parkinson's. Tanner said, "For the first time today, we can say that for people with Parkinson's disease diagnosed after age 50, it's most commonly caused by environmental factors."[80]

However, for those who develop Parkinson's before fifty, genetic factors appear to be more important.[81] The younger the age at onset, the greater the likelihood that genetics play a role.

Today, about 10% of people living with Parkinson's are thought to have an underlying genetic cause as the principal explanation for their disease.[82] While this is a small proportion, these genetic causes help scientists understand how Parkinson's develops and provide possible targets for new treatments.

As with many diseases, most cases of Parkinson's are likely due to a combination of environmental and genetic factors.[83] For example, not everyone exposed to paraquat even in high doses (e.g., farmers working with it for many years) develops Parkinson's disease. Most do not. Just as 85 to 90% of smokers do not get lung cancer.[84] For both diseases, other factors—some genetic—must be at play.[85]

Similarly, most genetic factors are insufficient to cause Parkinson's disease. For example, mutations in the LRRK2 gene do not always lead to the disease.[86]

So, scientists have looked for links between environmental and genetic factors and are starting to find them. In 2013, researchers found that the nerve cells of people with a mutation in the alpha-synuclein gene are more likely to die if they are exposed to certain pesticides such as paraquat and rotenone—even at levels considered safe.[87] And animal studies have also found that the same interaction can lead to the loss of dopamine-producing nerve cells.[88] Other studies have found similar results with LRRK2 mutations.[89]

Two hundred years after Parkinson's essay, researchers have identified many causes of Parkinson's, but more of them remain to be found. Like cancer, Parkinson's is not one disease but rather a collection of many with different contributing factors.

The late Dr. William Weiner, a Parkinson's expert at the University of Maryland, wrote in 2008 that "there is no single Parkinson disease and…there never has been." To reflect the multiple causes of the disease, including environmental ones, Weiner suggested the term "Parkinson diseases."[90] These different versions of the disease can have their own causes, symptoms, rates of progression, and, potentially, treatments.[91]

AN ASTONISHING NEW HYPOTHESIS

At the end of his 1817 essay, Dr. Parkinson urged other researchers to "humanely employ anatomical examination in detecting the causes and nature of . . . this malady." By doing so, he said its "real nature," treatment, or cure could be identified.[92]

Almost two centuries later, scientists took up this challenge. Two anatomists, Dr. Heiko Braak and his late wife, Dr. Eva Braak, examined hundreds of brains. In 2003, the Braaks and their colleagues proposed stages for the disease and a startling new hypothesis: Parkinson's does not begin in the brain.[93]

The researchers asked whether the disease might be caused by a yet unidentified pathogen, a pathogen being anything that causes disease. It could be a virus or a bacterium or some other external factor. We know that one cause for the misfolding of the alpha-synuclein protein is an inherited genetic mutation. But, Braak wondered, could some other pathogen be entering the body through the nose or gut, triggering alpha-synuclein to misfold, spreading to the brain, and causing the disease?[94]

FIGURE 5. Photograph of Dr. Heiko Braak and his wife, Dr. Kelly Del Tredici. *Courtesy of Dr. Heiko Braak.*

Pesticides and other chemicals in the environment or infectious particles are possible candidates. They can be inhaled through the nose or ingested through the digestive tract and can cause alpha-synuclein to misfold.[95]

Compelling evidence supports the hypothesis. Braak and his second wife, Dr. Kelly Del Tredici (**Figure 5**), have provided additional support for the contention.[96] Lewy bodies—those sacks of misfolded alpha-synuclein proteins—are found first not in the brain but in the nerves responsible for smell and gut motility, which reside in the nose and the intestines. The nose and the intestine may be the two points of entry for this neurological disease.

The nerve responsible for moving food through the gut is called the vagus nerve. This nerve begins near the bottom of the brain, in the brain stem, and travels to many destinations in addition to the gut, including the heart and lungs where it helps control heart rate and breathing.

For Parkinson's disease, the vagus nerve may be the road that enables the Braaks' "unidentified pathogen" to travel from the gut up to the brain, causing the misfolding of proteins as it goes.[97] From the vagus nerve, the pathology of Parkinson's—the mark of the disease—spreads upward like "a falling row of dominos" to higher brain centers that control movement and thinking.[98] Recent research confirms that the pathology of Parkinson's can indeed spread from nerve cell to nerve cell—the misfolding appears to be transmissible.[99]

The spread of Lewy bodies mirrors the development of symptoms.[100] The misfolded proteins that show up first in the nose and gut coincide with some of the first symptoms of Parkinson's—the loss of smell and constipation. These symptoms may begin years or decades before the classical movement symptoms develop.[101] The movement symptoms, such as tremors, occur when the disease has reached the substantia nigra in the brain. But the disease does not stop there. In the late stages of the disease, the Lewy bodies spread further to the outermost portions of the brain where they can cause dementia and hallucinations.[102]

The Braak hypothesis may indeed be correct. Parkinson's disease may have its origins outside the brain—and the body.[103]

Over the two centuries since the seminal description of Parkinson's disease, our appreciation of this malady has broadened. We now better understand the contributions of the environment, genes, and their interactions just in time to confront a pandemic of the disease.

2

A MAN-MADE PANDEMIC

How Chemicals Have Fueled the Onslaught

Could [Parkinson's disease] be a true man-made disease?

—Dr. William Langston, who identified MPTP
as a cause of parkinsonism, in 1997[1]

IN 1961, NEUROLOGISTS FROM ALL OVER THE COUNTRY GATH-
ered in Atlantic City, New Jersey, for the eighty-sixth annual meeting
of the American Neurological Association. They caught up with old
friends, exchanged gossip on the boardwalk, and heard an intriguing
new idea from two Harvard neurologists, Drs. David Poskanzer and
Robert Schwab. They argued that Parkinson's disease would disappear "as a major clinical entity by 1980."[2]

The roots of their claim lay in Vienna during World War I. An Austrian pilot named Constantin von Economo, who had been stationed
on the Russian front, returned to the city to resume his career as a neurologist. It was 1916, and his country needed him to care for wounded
soldiers.[3]

In addition to tending to veterans with head injuries, von Economo
saw patients with a strange new disease. The illness, which he described
as a "sleeping sickness," would strike individuals out of the blue. He
observed that people were "falling asleep while eating or working . . .
frequently in a most uncomfortable position."[4] After this abrupt onset

of sleep, a headache, nausea, and fever followed. Many would go into a coma and die.

The sleeping sickness spread throughout Europe and North America and affected about 1 million people worldwide between 1915 and 1926.[5] Then it vanished. By 1928, there were no new cases of the mysterious illness. Since then, only rare incidences have been reported.[6]

Those who recovered often developed different symptoms months or even years later. These included slow movements, stiffness, and tremors. They had what looked like Parkinson's disease.[7] The only differences were that these individuals were young and that a preceding infection had triggered the disease. Some were teenagers.[8] For decades, they would remain in physically frozen states, unable to move or communicate.

Dr. Oliver Sacks, the neurologist and author, described many of these patients in his classic book *Awakenings*. They had endured von Economo's sleeping sickness only to develop profound parkinsonism years later. He wrote, "Stares of immobility and arrest ... started to roll in a great sluggish, torpid tide over many of the survivors."[9]

The patients were "as insubstantial as ghosts, and as passive as zombies," Sacks wrote. They "were put away in chronic hospitals, nursing homes, lunatic asylums, or special colonies. [They] were totally forgotten. . . . And yet some lived on."[10]

Sacks was working at a psychiatric hospital in the Bronx in the 1960s when he first encountered Leonard Lowe. At forty-six years old, Lowe was mute and frozen except for tiny movements of his right hand. Lowe had developed the first sign of parkinsonism when he was a teenager. "His right hand started to become stiff, weak, pale, and shrunken," according to Sacks. The symptoms slowly progressed, but Lowe, an avid reader, was still able to graduate from Harvard with honors. Later, when he was working toward his PhD, "his disability became so severe as to bring his studies to a halt."[11] Lowe and patients like him were "awaiting an awakening."[12]

That awakening came in March 1969. Two weeks after Sacks started Lowe on levodopa (**Box A**), "a sudden 'conversion' took place.

BOX A. WHAT IS LEVODOPA?

Levodopa is the most effective medication for Parkinson's. In the brain, it is converted into dopamine, which is deficient in people with the disease. Like all drugs, however, it has side effects. Over time, high doses can trigger involuntary movements. These are often writhing or dance-like motions that can make it difficult for someone to sit or stand still. Levodopa and similar drugs can also cause debilitating impulsive behaviors.

The rigidity vanished from all his limbs, and he felt filled with an excess of energy and power. [He] became able to write and type once again, to rise from his chair, to walk with some assistance, and to speak in a loud and clear voice. . . . He enjoyed a mobility, a health, and a happiness which he had not known in thirty years."

Unfortunately for Lowe, levodopa's benefits were transient. The drug caused involuntary movements and aggressive behavior that led Sacks to stop the medication. Lowe, who would later be depicted by Robert De Niro in the movie *Awakenings*, never recovered. He died in 1981.

In the 1950s and 1960s, Poskanzer and Schwab, the neurologists who caused a stir at the meeting in Atlantic City, were also caring for individuals like Lowe. They believed that most cases of Parkinson's disease were due to the sleeping sickness. They thought that as these individuals died off, so, too, would the disease.[13]

Poskanzer was so certain of his view that in a 1974 *Time* magazine article, he said, "I offer a bottle of Scotch to any doctor in the U.S. who can send me a report of a clearly diagnosed case of Parkinson's in a patient born since 1931."[14] Of his bet, Dr. Poskanzer reported, "So far it's cost me 14 bottles—just 14 of these younger patients identified." The article in *Time* concluded, "[There] should be many fewer such patients [with Parkinson's disease] in the future—provided, of course, that Poskanzer wins his bet."

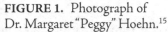

FIGURE 1. Photograph of Dr. Margaret "Peggy" Hoehn.[15]

Poskanzer, of course, lost. While most of his contemporaries had supported his and Schwab's contention, a few did not.[16] Dr. Margaret "Peggy" Hoehn was a pioneering neurologist at a time when few women were even admitted to medical school (**Figure 1**). Born in San Francisco in 1930 and trained in Canada and then at Queen's Square in London, Hoehn would become an associate professor of neurology at Columbia University and one of the world's leading authorities on Parkinson's disease.

Two years after Poskanzer's bold bet in *Time*, Hoehn made a distinction between the parkinsonism due to the sleeping sickness that von Economo described and the classical Parkinson's disease that James Parkinson had encountered.[17] The former was a consequence of a mysterious illness that had since disappeared. The second was rising in frequency. Even before Poskanzer's wager, Hoehn had demonstrated that the parkinsonism due to sleeping sickness was relatively rare. And as we know, far from disappearing, the number of people with Parkinson's has soared.

Over the last twenty-five years, Parkinson's disease is the only neurological disorder whose burden as measured by deaths, disability, and number affected have all increased even after adjusting for age.[18]

And the estimates are almost certainly low due to underreporting and missed or delayed diagnoses.[19]

THE ENGINE OF A SPREADING DISEASE

Dr. Parkinson wrote his 1817 essay in London, so it is linked in time and place to the height of the Industrial Revolution. The spread of the disease has closely tracked the growth of industrialization.

The new chemicals that this era has wrought are likely driving the spike in Parkinson's. Dozens of studies—both of humans and of animals in the lab—have linked the two. The list of known hazardous substances is extensive and goes beyond the pesticides paraquat, rotenone, and Agent Orange. It includes certain solvents, air pollution, and some metals, such as manganese used in welding.[20] Almost all of us have been, are, and will be exposed to these risks. Unless we want more of us to develop Parkinson's, we will have to change our practices.

The use of synthetic pesticides (**Box B**) began in earnest after World War II. By 1990, production soared to over 3 million tons per year, or more than a pound per person.[21] Between 1990 and 2016, the amount of pesticides increased by another 70%.[22] Over that same period, China's annual use of pesticides doubled from 0.8 million tons to 1.8 million. Each year, the country uses more than 2.5 pounds of pesticide for every person.[23]

Pesticides have improved crop yields and lowered costs, and many, if not most, are not linked to Parkinson's.[24] However, some of the most

---

BOX B. WHAT IS A PESTICIDE?

A pesticide is a chemical that is used to kill unwanted insects, animals, fungi, or plants. Insecticides, fungicides, and herbicides are all considered pesticides. Some of them (e.g., rotenone) are produced by plants to protect themselves. Others, called synthetic pesticides, are created by humans.

BOX C. WHAT IS A SOLVENT?

A solvent is a liquid that is used to dissolve another substance. Water is an example of a harmless one. It can dissolve many things—sugar, salt, and instant coffee, to name a few. Other common household solvents include nail polish remover and paint thinner. Industrial solvents are used in cosmetics to dissolve pigments, in dry cleaning to remove spots, and by the automotive industry to remove grease.

---

frequently used pesticides are associated with an increased risk of the disease. Given that there are pesticides that don't pose a Parkinson's risk, we can and should eliminate the ones that are known to be toxic to dopamine-producing brain cells.

Pesticides, of course, are not the only industrial product whose use has increased. Solvents, which are used to dissolve other substances (**Box C**), arrived in the latter half of the nineteenth century from the coal and tar industry.[25] Since then, these chemicals have evolved countless applications in consumer and industrial products, including in cosmetics, cleaning solutions, paints, pharmaceutical products, and automobile production.[26] By one estimate, 8% of the working population regularly uses solvents.[27] Almost all of us are exposed to them at home in the skin products that we use, the cleaning agents in our cabinets, the paints in our garages, and even the pills that we take.

Trichloroethylene, which is linked to Parkinson's disease, is one of the most common industrial solvents. Its production is on the rise globally, especially in China.[28]

Researchers have also found that air pollution increases the risk of Parkinson's.[29] Very small inhaled toxic particles can bypass the brain's normal protective mechanisms, injuring it directly. And air pollution has increased exponentially worldwide in step with global industrialization. The toxic smog in China's rapidly industrializing cities is comparable to the London fogs of the early Industrial Revolution.

Not surprisingly, the rates of Parkinson's have been increasing the most in industrializing countries. Over the last twenty-five years the prevalence rates for Parkinson's, adjusted for age, increased by 22% for the world, by 30% for India, and by 116% for China.[30]

The burden of Parkinson's falls more often on men, who are more likely to work in occupations that expose them to industrial products linked to the disease. In the United States, for example, men make up 75% of farmers, 80% of metal and plastic laborers, 90% of chemical workers, 91% of painters, 96% of welders, and 97% of pest control workers.[31] Men also have a 40% greater risk of developing Parkinson's than women.[32]

THE AGE FACTOR

One of the greatest human accomplishments of the twentieth century was the doubling of life expectancy.[33] In 1900, the average life span globally was just thirty-one years; by 2000, it was sixty-six.[34] The result is that the number of people over age sixty-five is increasing (**Figure 2**).

But as we age, the likelihood that many of us will develop Parkinson's increases.[35] Aging itself, though, is not likely the cause of the

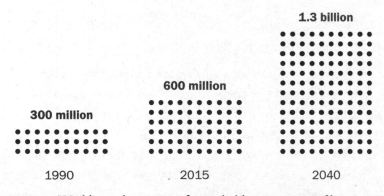

FIGURE 2. World population sixty-five and older, 1990–2040.[36]

disease. Rather, living longer allows time for nerve cell loss to occur and, therefore, for Parkinson's to develop.[37]

The environmental and genetic factors that contribute to Parkinson's require time for the damage to become apparent. The actual onset of the disease likely begins twenty or more years before symptoms like tremors appear.[38] During this period the disease may be spreading from the gut and nose to lower and then higher areas of the brain. As it stealthily takes hold and time passes, more nerve cells die. It's only when about 60% of nerve cells are gone that the typical features of Parkinson's finally show up.

And so, as the population of older people increases, so too will the number of individuals who experience Parkinson's disease. In fact, a person's risk rises sharply with age. Beginning in a person's forties, the risk of Parkinson's roughly triples with each passing decade.[39] In 2019 alone, 60,000 individuals, or over 1,000 Americans per week, were diagnosed with Parkinson's.

And we will keep living longer and longer. Recent headlines have highlighted small declines in life expectancy in the United States due to suicides and the opioid epidemic, but the long-term trend remains on an upward trajectory.[40] By 2030, there will be a 10% higher risk of a forty-five-year-old individual eventually developing Parkinson's disease than there is today.[41] With increasing longevity, more of us will face Parkinson's.

THE SMOKING PARADOX

Smoking's link to lung cancer provides a model for understanding the connection between environmental risk factors and Parkinson's disease. Like Parkinson's, lung cancer was once very rare.[42] Before the introduction of cigarettes, lung cancer was so uncommon that, according to Robert Proctor, a historian at Stanford University, "doctors took special notice when confronted with a case, thinking it a once-in-a-lifetime oddity."[43] With the industrial production and mass marketing of cigarettes in the late 1800s, however, lung cancer rates soon swelled.[44]

At the end of World War II, Britain had soaring rates of lung cancer, and no one knew why. Some speculated that dust from tarred roads might be the cause; others blamed poison gas from World War I.[45] To answer the question, Drs. Richard Doll, a physician and epidemiologist, and Bradford Hill, a statistician, conducted a simple study in 1951. They surveyed nearly 60,000 British physicians. They asked participants for their names, ages, and addresses. They also asked the doctors about their smoking history. The researchers then recorded the number and causes of deaths that subsequently developed in this group of physicians. They found a steadily rising mortality from lung cancer as the amount of smoking increased.[46] Further studies supported these results, leading to the conclusion that smoking caused lung cancer.[47]

With lung cancer, as with Parkinson's, the environmental risk has to occur over many years. Individuals who smoke for only brief periods (years and not decades) have a much lower risk of lung cancer.[48] In both diseases a lag is present between exposure and the development of disease.

In addition, the risk of lung cancer drops after people stop smoking.[49] This is the kind of opportunity that we should seek out in the case of Parkinson's—reducing our future exposure to the environmental risks linked to the disease.

Smoking has a counterintuitive connection to Parkinson's too. It actually *decreases* the risk of getting the disease—by an astonishing 40%, according to numerous studies.[50] The reason that smoking may lower the risk has yet to be determined. Some studies suggest that nicotine may protect nerve cells; others indicate that smoking may increase the breakdown of environmental toxins.[51]

Smoking might also confer its protective benefit by way of the nose or the gut.[52] It may block or otherwise interfere with the entry of the external factors that cause Parkinson's. Fumes from smoking are, of course, inhaled, and this changes the covering of the nasal passages and the local immune response.[53]

Smoking also changes the gut. The gut microbiome—the community of bacteria that live in our intestines—is remarkable. Over 100 trillion bacteria, far more than the number of human cells in the body, call us their home.[54]

Recent research has demonstrated that smoking and Parkinson's may affect the gut microbiome.[55] For example, smoking increases the population of certain bacteria that may boost the barrier function of the gut.[56] This barrier, the lining of our intestines, is the major interface between the environment—what we ingest—and ourselves. Its ability to keep out harmful substances is key.

So smokers' elevated levels of bacteria may help barrier function, thus potentially protecting them from toxins that up Parkinson's risk. Or perhaps their robust numbers of certain kinds of bacteria are keeping other kinds of harmful bacteria in check—bacteria that could be a factor in the development of Parkinson's. In a 2016 study, gut bacteria were shown to contribute to the development of the disease.[57] When genetically altered mice received antibiotics to kill off the bacteria in their guts, the pathology of Parkinson's was reduced.

Even more remarkably, this contribution from gut bacteria may be transmissible. When mice were given fecal material (which has numerous gut bacteria) from people who didn't have Parkinson's, the motor function of the mice was unchanged. However, when the mice received fecal material from people with Parkinson's, their motor function worsened. These results suggest that the gut bacteria may be influencing the brain. More research is required to explore the emerging gut-brain axis of disease, including the surreal possibility that poop transplantations may one day treat Parkinson's disease.[58]

Regardless of what the potential benefits of smoking might be, no one should smoke—or continue to smoke—to reduce the risk of Parkinson's. A causal relationship to the disease has not been proven. And the negative effects of smoking, including losing, on average, a decade of life, far outweigh any potential upside related to Parkinson's.[59]

Taken together, the increases in industrialization and aging and declines in smoking are likely to raise the future burden of Parkinson's far above the 12.9 million that we previously estimated. Future projections that account for these factors could lead to Parkinson's affecting 17.5 million people by 2040.[60] A far cry from the twenty-two who died of the condition in 1855 in England.[61]

A NEW KIND OF PANDEMIC

"Epidemiology is the study of what 'comes upon' groups of people," according to the late population health expert Dr. Abel Omran.[62] For most of human history, what has come upon people are famines and infectious diseases, sometimes in the form of a pandemic. As opposed to epidemics, which are limited to a given community, pandemics cover large geographic areas.[63] The black death killed up to 200 million people throughout Europe and Asia in the fourteenth century.[64] The influenza pandemic of 1918 killed between 50 million and 100 million people worldwide and led to an unprecedented drop in global life expectancy.[65]

Thanks to public health advances, infectious diseases are no longer the leading source of death and disability. The new leader of the pack is chronic conditions.[66] In his landmark 1971 paper on population change, Omran called our current era the "Age of Degenerative and Man-Made Diseases."[67] Among these are cancer, diabetes, heart disease, and Parkinson's.[68]

This transition has led some to expand the definition of "pandemic" to include diseases that are not infectious. The new carriers, or "vectors," of these diseases are not bacteria or viruses but urbanization, population aging, globalization, and the widespread availability of unhealthy products.[69]

Parkinson's disease satisfies many of the criteria of a pandemic (**Table 1**).[70] The disease is found everywhere in the world.[71] In almost every region, the rate of Parkinson's is increasing.[72] In addition, as with infectious pandemics, Parkinson's disease appears to be spreading and growing as industrialization expands.

CHARACTERISTIC	PARKINSON'S DISEASE
Geographic extension	Parkinson's affects individuals throughout the world.
Disease movement	The disease is spreading around the globe as populations age and countries industrialize.
Explosive rates	Parkinson's disease is the fastest-growing neurological disorder in the world, and its rates are increasing in almost every region of the world.
Minimal rates of immunity	No one (according to what we know now) is immune to Parkinson's disease. It affects men and women, the elderly as well as younger people, of all races and backgrounds.
Novelty	It was first clearly described only two hundred years ago and was likely very rare before then.
Infectiousness and contagiousness	The disease is not known to be infectious and is not transmissible from one individual to another. However, the pathology does spread within humans from nerve cell to nerve cell.
Severity	It is a feared and debilitating condition and ranks among the most devastating of all disorders.[73]

TABLE 1. Characteristics of a pandemic.

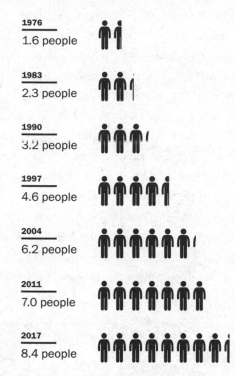

1976
1.6 people

1983
2.3 people

1990
3.2 people

1997
4.6 people

2004
6.2 people

2011
7.0 people

2017
8.4 people

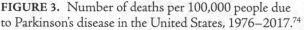

FIGURE 3. Number of deaths per 100,000 people due
to Parkinson's disease in the United States, 1976–2017.[74]

China is one extreme example. As the country experiences rapid
economic growth, it also has increased its use of pesticides and industrial solvents and has incredibly poor air quality.[75] Not surprisingly,
China also has the fastest-rising rate of Parkinson's in the world.[76]

As with other pandemics, the disease is severely debilitating. Over
its average fifteen-year course, Parkinson's leads to progressive loss
of independence and often nursing home care.[77] Although improved
therapies have increased survival, it remains a deadly condition. Death
rates from the disease are climbing in the United States (**Figure 3**),
where it is now the fourteenth-leading cause of death.[78]

Poskanzer and Schwab were unfortunately very wrong to predict
an imminent end to this formidable condition. A wave of Parkinson's

disease is upon us. And it affects everyone **(Figure 4)**: Democrats (Reverend Jesse Jackson) and Republicans (Senator Johnny Isakson); Catholics (Pope John Paul II) and Protestants (Reverend Billy Graham); capitalists (Jonathan Silverstein) and communists (Deng Xiaoping); activists (Walter Sisulu), actors (Alan Alda), actresses (Deborah Kerr), astronauts (Rich Clifford), and attorneys (Janet Reno); boxers (Muhammad Ali), cyclists (Davis Phinney), and runners (Sir Roger Bannister); baseball (Kirk Gibson), basketball (Brian Grant), football (Forrest Gregg), and hockey (Nathan Dempsey) players; journalists (Michael Kinsley) and photographers (Margaret Bourke-White); singers (Linda Ronstadt) and songwriters (Neil Diamond).

Humans have successfully confronted other diseases in the past—when they did not know what was coming. Parkinson's is different. We know it is bearing down on us. So we have a chance to learn from previous fights and prepare.

FIGURE 4. Faces of Parkinson's disease. *Courtesy of Dr. Bob Dein.*

3

VANQUISHING INDIFFERENCE

Lessons from the Fights Against Polio,
HIV/AIDS, and Breast Cancer

Until we get our acts together, we are as good as dead.

—Larry Kramer, playwright and AIDS activist[1]

ON THURSDAY, MAY 6, 1954, ROGER BANNISTER, A TWENTY-five-year-old medical student, finished rounds at St. Mary's Hospital in London. He then boarded a train to Oxford, lunched on a ham and cheese salad at a friend's house, and headed over to the Iffley Road track. At 4:30 p.m. he met his fellow amateur all-star teammates to race against the University of Oxford.[2] There, on a windy and rain-sodden day, he decided to attempt the unprecedented—to run a mile in less than four minutes. "I felt if I didn't take the opportunity when the chance presented itself," Bannister said, "I might never forgive myself."[3]

In front of 1,200 people in overcoats who had assembled to watch the meet on the quarter-mile track, he opened the race fast. Paced by two of his teammates, Bannister hit his stride. "I was relaxing so much that my mind seemed almost detached from my body," he said.[4] With rising applause from the crowd, momentum was building.

He ran the first three laps in three minutes, one second. It was fast but not fast enough, and Bannister knew it. He would have to find new speed on his last lap. He described the home stretch: "Time seemed to stand still. The only reality was the next 200 yards of track under my feet. I felt at that moment that it was my chance to do one thing supremely well.... My body must have exhausted its energy, but it still went on running just the same. . . . With five yards to go, the finishing line seemed almost to recede. Those last few seconds seemed an eternity. The faint line of the finishing tape stood ahead as a haven of peace after the struggle."[5]

After crossing the finish line, Bannister collapsed almost unconscious into the arms of racing staff. The crowd became silent. Did he break the four-minute barrier? The answer came from the track's announcer. Over a crackling speaker, he said, "Ladies and gentlemen, here is the result of Event No. 9, the one mile. First, No. 41, R.G. Bannister, Amateur Athletic Association and formerly of Exeter and Merton Colleges, Oxford, with a time that is a new meeting and track record and which, subject to ratification, will be a new English native, British national, British all-comers, European, British Empire and world's record. The time was 3 . . ." The ecstatic crowd drowned out the rest.[6]

Bannister was now the fastest runner in the world. But later that same year, at the peak of his racing career, Bannister retired to chase a new goal. He started training as a neurologist. Over the next forty years, he edited a neurology textbook, published important academic papers, and cared for countless patients, including many with Parkinson's.[7] In 1975, Bannister was knighted by Queen Elizabeth II.

In 2011, the former runner started having trouble walking. He was diagnosed with Parkinson's disease. Seven years later, at age eighty-eight, Sir Roger Bannister passed away peacefully in Oxford, England.[8]

During his lifetime, Bannister witnessed society confront and slow at least three pandemics: polio, HIV, and breast cancer. The lessons from these efforts can guide our response to the wave of Parkinson's that not even the great Roger Bannister could outrun.

POLIO: HALTING A PANDEMIC

Polio is an infectious disease caused by the polio virus. In the setting of poor hygiene, it is transmitted by mouth via contaminated water or food. It can result in weakness and even difficulty breathing. In the early twentieth century, the disease disproportionately affected children and terrorized families. That all began to change a hundred years ago.

In 1921, Franklin Delano Roosevelt—who had just lost his bid to become vice president—attended a Boy Scouts festival near his family estate in Hyde Park, New York. There he was photographed walking unassisted for the last time.[9] Roosevelt then traveled to his family's home off the coast of New Brunswick, Canada, where he sailed, helped put out a brush fire, raced his children, swam across a freshwater lake, dipped in the freezing waters of a local bay, and then developed a strange feeling of "numbness, deep muscle ache, and frightening chills."[10] Doctors later diagnosed him with polio.[11] Although some recent reports question the accuracy of the diagnosis, his ensuing paralysis helped define the future president and the fate of the disease.[12]

While his paralysis was hidden from the public, Roosevelt was focused behind the scenes on regaining strength and ending polio. Heartened by the buoyancy of the water in Warm Springs, Georgia, he bought the resort there and converted it into a foundation to help victims of polio. Later, in 1938, as president, Roosevelt announced the formation of the National Foundation for Infantile Paralysis. The group become the largest voluntary health organization of all time.[13] In 1954, for example, it raised more money than the American Cancer Society, American Heart Association, and National Tuberculosis Association combined.[14]

The foundation revolutionized philanthropy through novel approaches to fund-raising, many of them led by the group's president, Basil O'Connor. Among the foundation's earliest efforts was lining up Hollywood celebrities, such as the entertainer Eddie Cantor, to support the cause. For a 1938 campaign, Cantor suggested a "March of

Dimes," a campaign that would involve asking people to send their dimes directly to the President at the White House.[15] Even though much of the foundation's staff opposed the idea, O'Connor asked President Roosevelt, who replied, "Go ahead."[16]

With the President's endorsement, Cantor used his popular radio show to launch the March of Dimes effort to "enable all persons, even the children, to show our [President] we are with him in this battle." Jack Benny, Bing Crosby, and the Lone Ranger soon joined the cause and made their own appeals to support the March of Dimes.

The White House expected a modest increase in mail. What arrived instead was a tsunami. Ira R. T. Smith, who worked in the White House mail room for more than fifty-two years, recalled, "Two days later, the roof fell in—on me. We had been handling 5,000 letters a day at that time. We got 30,000 on the day the March of Dimes began. We got 50,000 the next day. We got 150,000 the third day. We kept on getting incredible numbers, and the Government of the United States darned near stopped functioning because we couldn't clear away enough dimes."[17]

In all, citizens mailed 2.7 million dimes, countless dollar bills, and numerous checks to the White House. The March of Dimes eventually became the official name of the foundation.[18]

The call for dimes was not the foundation's only campaign. The group collected money in movie theaters, hosted fashion shows to raise more cash, and created the first "poster child," who would become the face of the illness on countless posters and pamphlets that blanketed the country in the decades to come.

The foundation also launched a first-of-its-kind Mothers' March on Polio to raise money. On a January evening in 1950, crowds of women, many from minority groups, took to the streets of Phoenix, Arizona. Leading up to that night, a local chapter of the foundation had placed newspaper, radio, and billboard ads and passed flyers out to children at school. The message was simple: "Turn on Your Porch Light! Help Fight Polio Tonight!" And so for one hour, volunteers went door-to-door, collecting donations from contributors who signaled their willingness to give by turning on their porch lights.[19] The city "came alive. Sirens wailed, car horns sounded, and searchlights swept the sky."[20]

The donations helped care for people with polio. From 1938 to 1955, the foundation spent approximately two-thirds of its total budget, or $233 million, on patients. Most of it went to pay the individual medical bills of those affected because in 1940 less than 10% of the country's population had health insurance.[21] The foundation's vision was that "no victim of infantile paralysis, regardless of age, race, creed or color, shall go without care for lack of money." To fulfill that promise, the foundation provided aid to more than 80% of US patients with polio.[22]

Even as its victims were being cared for, the pandemic continued its sweep. By the 1950s, polio was generating so much anxiety that people started staying away from gathering places that gave the virus more opportunity to spread. Churches and swimming pools closed. In 1952, Americans feared polio more than anything except nuclear war.[23] That year, polio had its worst outbreak in the United States, affecting 58,000 individuals, paralyzing 21,000, and killing 3,000.[24]

Treatment options were few. In the early stages of the disease, when muscles in the chest became paralyzed, iron lungs—giant metal cylinders in which patients had to lie down—assisted with breathing by regulating air pressure. At the height of the pandemic, dozens of children in iron lungs filled hospitals throughout the country.[25] They spent day after day lying encased in these tubes with only their heads exposed. A mirror above their eyes allowed them to see more of their surroundings. After several weeks most would recover enough to breathe on their own, but some people ended up relying on iron lungs for years.[26] Obviously, iron lungs were not the answer to polio.

Vaccines, which had recently become available for tetanus, diphtheria, and pertussis, offered hope for a better solution. To develop one, researchers first had to identify the cause or causes of polio. They found that multiple strains of a polio virus were responsible for the disease. With this information, Drs. Jonas Salk and Albert Sabin led efforts to make a polio vaccine a reality.

By 1953, Salk had developed a vaccine containing the dead virus, which was ready to be tested widely. More than 1.5 million children were enrolled in what was "among the largest and most publicized clinical trials ever undertaken."[27] In 1954, over 600,000 children from across the country, from rural and urban areas, African American and Caucasian, wealthy and poor, participated in the trials. The vaccine worked, and in 1955, it was adopted widely in the United States.

In five years, polio cases dropped tenfold from the peak of 58,000 in 1952 to 5,600 in 1957. When famed reporter Edward Murrow asked Salk who owned the patent on the polio vaccine, he replied, "Well, the people, I would say. There is no patent. Could you patent the sun?"[28]

In 1961, six years after Salk's injection, Sabin developed a weakened live virus vaccine that could be swallowed and thus administered more easily. This new vaccine enabled mass-immunization efforts world-wide that to this day have reached hundreds of millions, if not billions, of children.[29] Thanks to generous philanthropic support, determined scientists, and broad societal engagement, polio today is close to join-ing smallpox in the dustbin of diseases that humans have eradicated from earth.[30]

HIV: A NEW BREED OF ADVOCACY

On July 3, 1981, the *New York Times* ran the headline "Rare Cancer Seen in 41 Homosexuals."[31] Eight of the men, according to the article, died less than twenty-four hours after they were diagnosed.

They had been infected with what researchers later identified as the previously unknown human immunodeficiency virus (HIV). Forty years after the polio epidemic, this new virus now threatened to kill millions around the world. It attacked the immune systems of those it infected. Over time, their immune cells became so damaged that it was difficult for them to fight off infections and cancer. This advanced stage of the disease is called acquired immunodeficiency syndrome, or AIDS.

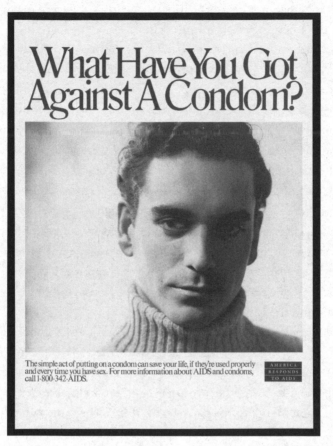

FIGURE 1. An example of AIDS educational materials, 1980s.

Yet, in just fifteen years, thanks to unprecedented activism and huge leaps in science, the fatal condition became a treatable one. And a cure is no longer inconceivable. In 2019, for the second time someone who had been infected with HIV was reported to be free of the virus.[32]

At its peak in 1997, 3.5 million people per year were diagnosed with HIV globally, and in 2005 AIDS was killing 2 million people annually.[33] In contrast to Roosevelt's leadership against polio, the federal response to HIV in the 1980s moved from ignorance to silence to blaming the victims. As just one example in 1982, President Ronald Reagan's press secretary joked about AIDS during a press briefing and said that he had not heard the president express concern or "anything" about the emerging public health crisis.[34]

Bold, determined activism rose to fill the void. An early and sustained focus of this engagement was prevention. When HIV was first reported, little was known about it other than its association with homosexual men in the United States. With no help from outside, the gay community sought to protect itself. Working with a few researchers, the community identified practices that were linked to the spread or transmission of the disease. Once sexual contact was identified as a risk, gay bathhouses were eventually closed, sexual practices changed, and abstinence increased. Education campaigns—posters adorned gay bars and city streets—led to the widespread use of condoms, which prevented transmission of not only HIV but also many other sexually transmitted diseases (**Figure 1**).[35] The gay community's efforts, radical at the time, were responsible for preventing the spread of HIV to millions around the world.

Today, prevention remains the cornerstone of the continued fight against HIV. The same medications that are used to treat someone with HIV can now be used to block its spread. These drugs help keep HIV from passing from mothers to their children and decrease the risk of transmission to uninfected partners by 96%.[36] The campaign to prevent HIV has been so effective that the number of new individuals infected with the virus has decreased for the last twenty years, deaths have fallen for the last ten, and the survival rate has increased for the past thirty.[37]

Early on, people with HIV/AIDS could not receive proper care because of who they were. Polio had primarily affected children. In the United States, HIV had mostly infected homosexual men. Even rising death tolls could not quell the rampant prejudice. During the early years of the epidemic, few doctors were willing to offer care, and even fewer were knowledgeable about the disease.[38] Many hospitals, including some of the most prestigious in the country, refused to admit those with HIV.[39]

The community was left on its own to suffer from and fight what was thought to be its disease. Its members also had to care for themselves. Buddies were assigned to everyone affected to "help them get to their appointments, pick up their prescriptions . . . walk their dog . . . do grocery shopping for them, [and] change their bed."[40]

The first treatment for AIDS was still years away. But patients were desperate to try anything. In response, AIDS activists ran their own pharmacies out of churches. They produced their own experimental therapies, imported antiviral drugs from other countries, and distributed a wide range of medications to their peers. They served tens of thousands of customers. For those who could not reach or afford a hospital, friends and family members offered up their own homes for the sick. In essence, the community created its own shadow health-care system.[41]

Although the AIDS community initially lacked a face for the disease, that would soon change. In 1985, Rock Hudson, beloved leading man of Hollywood and a friend of President Ronald and First Lady Nancy Reagan, revealed to a shocked public that he had AIDS. The fact that Hudson, the epitome of masculinity, could suffer such a fate made the disease far more socially acceptable.

One year later, thirteen-year-old Ryan White helped people realize that anyone could become infected. But he still faced vicious ignorance. He contracted HIV from a blood transfusion to treat his hemophilia, but he was barred from attending his middle school and faced immense prejudice. "People said that he had to be gay, that he had to have done something bad or wrong, or he wouldn't have had it," his mother said.[42]

The family eventually moved to another town in Indiana where Ryan was welcomed into the local school. His case led to far greater public and federal support for research and care, including legislation signed by President George H. W. Bush. The Ryan White Act, which provided $220 million to increase access and improve treatment for people with HIV/AIDS, was passed in 1990, several months after Ryan died. He was eighteen.[43]

But widespread indifference is not undone so easily. In 1987, Cleve Jones was doing his part. Jones was a protégé of Harvey Milk, the pioneering San Francisco politician who had been assassinated nine years earlier. At a candlelight march in San Francisco, Jones asked participants to write the names of friends and loved ones who had died of AIDS on cardboard placards. Many of those named had not received a proper memorial service. Some people had been abandoned by their families; for others, funeral homes had refused to offer services. Jones took the signs, taped them together, and hung them from the San Francisco Federal Building. The prototype for the AIDS Memorial Quilt was born.[44]

Four months later, on October 11, 1987, 500,000 people gathered in Washington, DC, to see the football-field-sized AIDS Memorial Quilt displayed on the National Mall. The quilt contained 1,920 fabric panels, each approximately the size of a grave, commemorating the life of someone who had died of AIDS.[45] David France, author of *How to Survive a Plague*, recounts the day:

> From blocks away we could hear the dull cadence of a lone female voice reciting the names of the men and women whose elegies were stitched into a cloth tribute called the AIDS Memorial Quilt. . . . We stood on the edge of the massive field listening to muffled sobs around us as a platoon of pallbearers, dressed uniformly in white, unfolded great colorful panels and slowly lifted them into their place on the grid. There, beneath the doleful roll call of names, they silently tied the panels together with ribbons. . . . The speaker paused after each name, and you could sometimes hear combustions of grief from their loved ones roll across the meadow.[46]

One year later, the quilt had grown to include over 8,000 panels and was displayed on the Ellipse in front of the White House. By 1996, nine years after Jones started assembling his memorial, the quilt covered the entire National Mall. Today, the quilt is shown at exhibits throughout the United States and remains the largest community art project in the world.[47]

The quilt debuted several months after the Food and Drug Administration (FDA) approved AZT, the first treatment for AIDS. It had been six years since that *New York Times* headline about the rare cancer in homosexual men, and millions were now infected. Patients—and the wider, terrified public—were desperate for some kind of therapy. The FDA, facing enormous pressure, had fast-tracked approval of the drug in a record twenty months. AZT was originally developed as a cancer drug, and in a rushed clinical trial, it reduced the risk of death from AIDS.[10]

When AZT came on the market, AIDS advocates accused the manufacturer of profiting off a vulnerable population. It cost $8,000 a year, the equivalent of $17,000 today.[49] AIDS disproportionately affected individuals with limited financial means. Many had no health insurance. Some who did frequently became too sick to work and consequently lost their income and health insurance when they were needed most. Activists wanted to ensure that the entire community, not just the wealthy or the well insured, would benefit from new therapies.

Peter Staley, an investment banker with HIV, had become involved in ACT UP, the radical protest organization that began in New York City and whose motto was "Silence = Death." To protest the cost of AZT, Staley and his fellow activists barricaded themselves inside the national headquarters of Burroughs Wellcome, the pharmaceutical company that made AZT. Then, months later, Staley and his friends dressed in suits and entered the New York Stock Exchange. There, they chained themselves to a balcony above the trading floor, unfurled a banner that read, "Sell Wellcome," and drowned out the opening bell with air horns. Within days, Wellcome had lowered AZT's price by 20%.[50]

AZT, however, was not enough. After an initial response, AZT lost its effectiveness. Additional treatments were needed.

The community quickly grasped that for new drugs to be developed, it would have to continue to pressure government agencies. In 1988, ACT UP orchestrated a carefully designed plan to occupy the FDA's headquarters. In the lead-up, the group prepared a forty-page handbook for anyone interested in participating, educated members on the regulatory process, and gave newspaper reporters and TV stations advanced notice of what was to come. When the appointed day arrived, activists stormed the FDA's building in Rockville, Maryland, shutting it down with hundreds of protestors.[51] Banners reading, "We Die—They Do Nothing," covered the campus and the building's facade, and according to the late writer and professor Douglas Crimp, "[Demonstrators] staged one piece of theater after another as the television cameras rolled."[52]

After the successful demonstration at the FDA, that federal agency, along with the National Institutes of Health (NIH), began listening to activists. Their input and thoughts on clinical trial design were solicited, followed, and incorporated into future studies.[53] NIH funding for HIV/AIDS research increased, and today the agency spends more on HIV annually—$3 billion—than on any other single condition.[54]

This activism, increased research funding, and eventually new therapies transformed the fate of those with HIV. The success of the HIV/AIDS movement offers a key lesson for drug development and access. The immense public health gains have not stifled innovation or profits. From 2000 to 2018, the FDA approved thirty-one drugs for HIV.[55] Along with these new therapies have come profits. During roughly the same period, pharmaceutical companies including Bristol-Myers Squibb, Gilead, and GlaxoSmithKline (with which Burroughs Wellcome merged) earned $200 billion in sales from HIV drugs. Profits quintupled from less than $5 billion to $25 billion.[56]

Five years into the epidemic, these successes were not even a dream. Some elected officials remained impervious to the ongoing tragedy. Senator Jesse Helms ardently opposed using federal funds to study HIV/AIDS. He voted against the Ryan White Act and refused to speak to Ryan's mother even when he was alone with her in an elevator. He claimed, "There is not one single case of AIDS in this country that cannot be traced in origin to sodomy."[57]

Peter Staley, the investment banker activist, had had enough.[58] He was upset that the senator had gotten a pass. Staley said, "One of the best tools an activist can use is humor. If you get folks laughing at your target's expense, you diminish his power. I wanted the country to have a good laugh at Helms' expense. I wanted his fellow Senators to have a little chuckle behind his back. And I wanted Senator Helms to realize that his free ride was up—if he hit us again, we'd hit back."[59] So Staley and seven fellow agitators from ACT UP came up with a plan to rectify the situation: put a giant condom over Senator Helms's home. After locating his house in Arlington, Virginia, and measuring its dimensions, the group sent the specifications to three companies in California that specialized in custom-made inflatables like those used in grand openings for shopping malls. They chose the company that offered the lowest price.[60]

The activists planned out every step of the process and assigned each team member a role, from operating an air blower on the ground to handling questions from the media. Then, after ensuring that the home was empty, they climbed up ladders to the senator's roof, hauling with them a large duffle bag. From it, they unfurled the bespoke contraceptive over the house. It read, "A condom to stop unsafe politics. Helms is deadlier than a virus."[61] A week later, the senator complained about the act on the Senate floor, but according to Staley, "Helms never proposed or passed another life-threatening AIDS amendment."[62]

The might of the AIDS activists and the change they wrought have reverberated far beyond the borders of the United States. Today, 79% of people worldwide who know that they are HIV-positive have access to effective therapies; 81% of people who receive treatment have undetectable levels of the virus in their blood.[63] The United Nations aims to ensure that, by 2020, 90% of the world's diagnosed population has access to antiretroviral treatment and for 90% of them to have suppressed the virus.[64] In forty years, advocates transformed an unknown, fatal illness into one of the most treated conditions in the world.

BREAST CANCER: ALLEVIATING STIGMA

Like HIV, breast cancer used to have an enormous stigma attached to it. In the face of a wave of breast cancer, a new form of activism, one that has found far greater social and even corporate acceptance, has emerged. The result is heightened awareness and support for a disease that less than fifty years ago was rarely discussed.

In the fall of 1974, seven weeks after the resignation of President Richard Nixon, First Lady Betty Ford broke the seal, telling the nation and the world that she had breast cancer. As Roosevelt had, the First Lady gave both a face and a voice to a modern epidemic.[65] Ford's disclosure was met with compassion, not scorn. More than 50,000 people, including many with breast cancer, mailed her letters.[66]

Speaking of her illness, Ford said, "There had been so much cover-up during Watergate that we wanted to be sure there would be no cover-up in the Ford administration."[67] In the wake of Ford's courage, Happy Rockefeller, wife of Vice President Nelson Rockefeller, soon sought testing and was found to have breast cancer. She too shared her diagnosis with the country. American women were encouraged to perform self-examinations and pursue diagnostic testing.

The impact of these two voices was huge. With screenings up, the actual number of new diagnoses of breast cancer in the United States soon increased 15% and was termed the "Betty Ford blip."[68] Ford said, "[The] fact that I was the wife of the President put it in headlines and brought before the public this particular experience I was going through. It made a lot of women realize that it could happen to them. I'm sure it saved at least one person—maybe more."[69]

Before the First Lady's public disclosure, women had already been working to raise awareness of breast cancer. In 1973, childhood movie star Shirley Temple Black became the first public figure to write about her experience with the disease. The title of her article for the women's magazine *McCall's* was "Don't Sit Home and Be Afraid."[70]

Even after Betty Ford's galvanizing admission, there was still work to be done. Dr. Audre Lorde, the poet and feminist, was diagnosed with breast cancer in 1978. She railed against the prevailing paternalism exemplified by her physician, who told her, "If you do not do exactly what I tell you to do right now without questions, you are going to die a horrible death." Following Lorde's mastectomy, another doctor repeatedly noted her "obese abdomen and remaining pendulous breast."[71] In her book *The Cancer Journals*, Lorde criticized oppressive social views of the disease, which had lingered. Silence on the subject was, as it would be for the AIDS activists, unacceptable, for, she wrote, "silence has never brought us anything of worth."[72]

When Nancy Brinker's sister, Susan, was dying of breast cancer in 1981, she decided she could not keep quiet either. Nancy promised her sister that she would do whatever she could to end the pain and hopelessness of having breast cancer. A year later, and two years prior to her own diagnosis of breast cancer, she founded the Susan G. Komen Breast Cancer Foundation.[73]

Today the foundation's signature fund-raising event, the Susan G. Komen Race for the Cure, draws nearly 1 million participants in 140 events over four continents.[74] Since 1982, the foundation has raised awareness and nearly $1 billion for breast cancer research.[75] Only the US federal government has spent more.

A decade later, Charlotte Haley, a sixty-eight-year-old housewife living in a sprawling suburb of Los Angeles, picked up her own megaphone. Both her older sister and her daughter had been diagnosed with breast cancer in the 1980s. Haley was frustrated by the lack of progress against the disease, especially the scarcity of funding for prevention.

So she started a grassroots campaign in her dining room. There she began making little loops out of peach-colored ribbons. She packaged five of them together and attached a postcard with a simple message: "The National Cancer Institute's annual budget is $1.8 billion, only 5% goes for cancer prevention. Help us wake up legislators and America by wearing this ribbon." She left them in doctors' offices, handed them out in grocery store parking lots, and mailed them to prominent women, including former First Ladies.[76] Her husband worked extra hours to pay for the photocopying, ribbons, and postage. She eventually made 40,000 of them, but Haley refused to accept financial support, returning any checks that she received and telling the donor to give the money to cancer research instead.[77]

Haley's efforts began to draw attention from the media, including the *Los Angeles Times*. Soon she received a call from the editor of *Self* magazine, who wanted to use the peach ribbons to promote breast cancer awareness in its pages. Working with the magazine, Estée Lauder also planned to distribute the ribbons at the company's cosmetic counters throughout the country. Haley said, "No, you're going to commercialize it. That's making money off of somebody else's pain and suffering, and I've been through that with my sister and daughter, and we just can't do that." Rebuffed, *Self* magazine consulted with its lawyers and considered its options. The publication settled on a different color: pink.[78] In 1992, Estée Lauder distributed 1.5 million pink ribbons along with instructions on how to perform a proper breast self-exam.[79]

Today, activists have delivered on their promise to change how we think about and treat breast cancer. The shame tied to the condition has been erased. The NIH spends over $700 million annually on breast cancer research, and new treatments are being developed.[80] Breast cancer diagnoses have plateaued, and five-year survival has increased to 90%.[81]

$$\emptyset \; \emptyset \; \emptyset$$

THE CAMPAIGNS AGAINST POLIO, HIV, AND BREAST CANCER HAVE many lessons for us as we take on Parkinson's. Polio was eradicated only with presidential leadership and extensive community engagement that included children. Creative and novel campaigns first funded care for those in need and then financed research to identify the causes of the disease. Effective treatments (in this case, vaccines) were then tested on unprecedented numbers of volunteers at risk of developing the disease.

For HIV, activism overcame ignorance, indifference, and prejudice. While the means were provocative and even disturbing, the ends were enduring. Those with the disease now have a voice in the conduct and funding of research, a voice that needs to be amplified. Bureaucracies are more accountable to their funders and beneficiaries, but that accountability needs to increase. Care and treatment were extended in large measure to all those with the disease, an effort that needs to expand. While more work remains, a global threat to public health was met when those affected refused to be silent.

Progress against breast cancer began when brave, high-profile women were willing to confront the social stigma associated with the disease and share their stories. This led to ordinary women taking extraordinary steps to raise awareness and funds for research into the disease. Those funds have facilitated early diagnosis and new treatments for the condition. More work remains, however, to prevent breast cancer.

In all cases, activists refused to accept that these diseases were inevitable. As we face the new challenge of Parkinson's, these activists have taught us how to overcome indifference and use our voices. Based on their experiences, we must form a "PACT" to end Parkinson's. This PACT will (1) *prevent* the disease, (2) *advocate* for policies and resources, (3) *care* for all affected, and (4) *treat* the condition with new and more effective therapies.

PART TWO

The PACT—Preventing, Advocating, Caring, and Treating

BEFORE IT STARTS

The Urgency of Banning Specific
Pesticides to Lower Our Risk

The farmers spoke of much illness among their families. In the town the doctors had become more and more puzzled by new kinds of sickness appearing among their patients.

—Rachel Carson, *Silent Spring*, 1962[1]

OVER THE PAST FORTY YEARS, FIVE MEMBERS OF TERRI Mc-Grath's extended family, including her beloved grandfather, developed Parkinson's disease. In 2005, McGrath, then a forty-nine-year-old special education teacher, noticed that when she walked, her left arm was not swinging and her left foot was dragging. Three years later, she became the sixth member of her family, and the first woman, to receive the same diagnosis.

In the 1920s, McGrath's grandfather, Alex Adent, and nine of his brothers and sisters emigrated from Lithuania to the United States. Most of the sisters became seamstresses, and most of the brothers, including McGrath's grandfather, became farmers in southwestern Michigan. Ever since then, McGrath's family life has centered on a farm in St. Joseph, a small town on the eastern shore of Lake Michigan. Her eighty-six-year-old mother spent her entire life within two miles of that farm.

As a child, McGrath was always playing outside with her grandparents. The family regularly used pesticides on its farm, including DDT, until it was banned in the United States in 1972.[2] When the older family members sprayed the chemicals, McGrath stayed inside because her grandmother did not want her to breathe them in. But McGrath would often slip outside soon after, while the pesticides still lingered in the air.[3] She, her siblings, and cousins would all pick apples, currants, and grapes that were covered with a white film. Only after her diagnosis did McGrath learn how dangerous some of those pesticides were and how some were linked to Parkinson's.[4]

McGrath loved growing up on that farm. Despite her diagnosis, she says that had she known about the harmful effects of pesticides, she is not certain she would have done anything differently. Her children, who now are all "super conscious" of what they eat, are surprised that their mother did not think more about the risk of pesticides.[5] McGrath says, "I just didn't think about [it]. I thought farming was fun."

McGrath no longer lives on the farm, but her uncle and cousin do. Both are embracing organic farming. McGrath herself remains active. Now retired, she still tutors, is planning a camping trip along the Natchez Trail in the southern United States, and is awaiting the birth of her ninth grandchild. Although she is doing well currently, could she and her family have avoided Parkinson's disease?

The answer is likely yes. Farmers who are exposed to certain pesticides have a higher risk of developing the disease.[6] In one study, the risk of developing the disease for farmers was 170% greater than that for nonfarmers.[7] And the longer the farmers worked with the pesticides, the greater their risk.[8]

Pesticide risks are not limited to farmers. People who simply live in rural areas have high rates of Parkinson's disease.[9] These individuals may be exposed to pesticides in the air that can drift into residential communities.[10] In addition, pesticides can contaminate groundwater or well water.[11] Private wells are often shallow and may be especially at risk for contamination from nearby pesticides.[12] Moreover, in the

United States, private well water is not subject to the same regulations as water that comes from public systems.[13]

Agricultural areas have the highest rates of Parkinson's. In Nebraska, the rates of the disease are two to four times higher in the state's rural, agricultural parts than in urban Omaha.[14] In Canada, investigators have found an almost perfect correlation between areas with the highest pesticide use and the highest rates of disease.[15] In France, rural areas have the highest rates of Parkinson's, as do the regions with the most vineyards, which often require intense pesticide use.[16]

And then there is the rest of us. We eat fruits, vegetables, nuts, and grains every day that have been doused in pesticides. What kind of risk are we all exposed to? We do not know. Assessing pesticide exposure from foods and documenting individuals' dietary habits over a lifetime—given that Parkinson's takes decades to unfold—are challenging. But this research is needed.

Until then, we are left with educated guesses. We do know that organic foods have much lower levels of pesticide residues than conventional choices.[17]

DDT ON THE FARM

The insecticide DDT was once considered a miracle. In the 1930s, the Swiss chemist Dr. Paul Hermann Müller was looking for a chemical that could kill insects that were destroying crops and spreading disease—without harming the plants. Müller, a nature lover, tested hundreds of chemicals before coating the inside of a glass box with DDT, a colorless, tasteless, and almost odorless nerve toxin.[18] He placed houseflies into the container, and they bit the dust. Müller had found his answer.[19]

During World War II, DDT, which kills a wide range of insects, slowed the spread of malaria and other diseases among Allied forces in Europe and in the South Pacific. Winston Churchill said in 1944, "The excellent DDT powder which had been fully experimented with and found to yield astonishing results will henceforth be used on a great scale by the British forces in Burma and by the American and Australian forces in the Pacific and India."[20] According to the historian Dr. James Whorton, "When the war ended, DDT was given a hero's welcome, hailed throughout the land as 'Killer of Killers' and 'the atomic bomb of the insect world.'"[21]

In 1948, Müller received the Nobel Prize for Medicine. The Nobel Committee said, "Without any doubt, the material [DDT] has already preserved the life and health of hundreds of thousands." Today, DDT is still used to help control the spread of malaria in parts of Africa.[22]

During the war, DDT was also used at home. Dr. Guy Wilcox's father supplied food for soldiers from his family's farm in the small hamlet of Sauquoit in upstate New York. Wilcox's father raised dairy cows and grew grain. To protect the oats and barley, they used pesticides, including DDT. The chemical was stored in an old barn where Wilcox spent many hours as a boy. The can of DDT was often covered in cobwebs that kept away flies but not his young hands. Wilcox would pry open the can, play with the powder, and often make a mess in the barn.

DDT was easy to use and inexpensive. Wilcox helped his dad mix it with fertilizer for the crops. They would then spread the mixture over the fields while undoubtedly breathing it in.

In 2008, Wilcox began dragging his right foot. He developed a tremor in his right hand and was diagnosed with Parkinson's. A decade later, the progression of his disease forced him to stop practicing medicine.

DDT was not the miracle chemical that Müller had envisioned. Beginning as early as the 1940s, its harmful effects on the health of wildlife, humans, and the environment were identified and detailed, including in Rachel Carson's 1962 book *Silent Spring*.[23] Even though it was banned half a century ago, DDT persists in the environment— and in our food supply. It becomes more concentrated as it makes its way up the chain to human consumption. The pesticide is then stored in our fatty tissues, as is the case with other animals.[24]

Because of the widespread use of DDT and related chemicals, some are detectable in nearly everyone.[25] In 2003 and 2004, more than thirty years after the insecticide was banned, the US Centers for Disease Control tested the blood of about 2,000 people ages twelve and older. The researchers were looking for DDT and its metabolite, or break-down product, DDE. The body converts chemicals in food and medications to often simpler molecules that are sometimes referred to as "metabolites." They found that "a small proportion of the population had measurable DDT [and] most of the [US] population had detectable DDE" in their blood.[26] For Parkinson's, what matters more are the concentrations of chemicals in the brain. Because DDT dissolves in fat, "levels of DDT or metabolites may occur in fatty tissues [such as the brain] at levels up to several hundred times that seen in the blood."[27]

AGENT ORANGE IN VIETNAM

Vietnam veterans and up to 4 million Vietnamese came into contact with Agent Orange during the war. Named for the color of the 208-liter barrels in which it was stored, Agent Orange was an herbicide mixture designed to kill vegetation and crops in the tropical forests so that aircraft fighters could spot the enemy below.

From 1965 to 1970, an estimated 45 million liters of Agent Orange were sprayed in Vietnam (**Figure 1**).[28] There has been no large-scale study of the effect of this exposure on the health of the Vietnamese or war veterans.[29] Smaller studies, however, have linked Agent Orange to many problems in these populations, including birth defects, cancer, and Parkinson's.[30] People who were exposed to Agent Orange during the war have been found to have a higher risk of developing the disease.[31] Studies looking at Korean veterans who were in Vietnam also show an association between Agent Orange exposure and subsequent development of Parkinson's. The evidence is sufficient that veterans who were exposed to Agent Orange and now have Parkinson's are eligible for disability compensation and health care from the US Department of Veterans Affairs.[32]

FIGURE 1. Planes spraying Agent Orange in Vietnam.

Richard Stewart, the son of a World War II veteran, is a seventy-one-year-old former Green Beret who has Parkinson's disease. At the time of the Vietnam War draft, Stewart was working for Eastman Kodak developing reconnaissance tools, so he was eligible for occupational deferment. But he chose to enlist and deployed to Vietnam in 1970. He later became a platoon leader in the US Army's famous 101st Airborne Division. He remembers being in and out of areas treated with Agent Orange. "Pretty nasty stuff," Stewart recalls.

Four decades later, he noticed that he was beginning to lose his sense of smell, and later he developed a tremor in his left hand. He was diagnosed with Parkinson's. Stewart had become a mathematics educator at age fifty-one, but he was forced to retire early due to progression of the disease.

Today, Stewart lives in upstate New York with his wife, a self-described "flower child who peacefully protested the war," who was ready to demonstrate again after her husband's diagnosis. Stewart remains physically active, walking 2.5 miles and doing two hundred push-ups a day. He is also an enthusiastic member of veterans' groups and says, "I only have Parkinson's disease. A lot of people are a lot worse off."

CONTAMINATED MILK

On March 20, 1982, a New York Times headline read, "Hawaii Recalls Pesticide-Laced Milk from Stores and Schools."[33] Farmers in Oahu had fed dairy cattle the tops of pineapple plants known as "green chop." The problem was that the leaves had been sprayed with heptachlor, a pesticide that had been banned by the Environmental Protection Agency (EPA). Heptachlor was considered a potential carcinogen, or cancer-causing chemical, and it persists in the environment.[34] In the 1960s, the Food and Drug Administration had set a zero tolerance level for heptachlor in foods.

In 1978, the Pineapple Growers Association of Hawaii and the state of Hawaii argued that heptachlor was needed to protect pineapples, the state's largest source of agricultural revenue, from ants. At the time heptachlor could be found in the fat cells of virtually every American.[35] The attorney representing the state of Hawaii and the pineapple growers said that his clients "do not concur or accept . . . that heptachlor use to control ants on pineapple poses a significant risk of exposure to man or the environment."[36] They were granted an exemption and started using the pesticide.

The exemption allowed heptachlor to be sprayed on pineapples but prohibited the use of green chop as feed within one year of spraying it with the chemical. Then, in 1982, it was discovered that some of the contaminated tops apparently had slipped into the feed before the year was up and the heptachlor had broken down at least partially.[37] Testing found that seven of the nineteen dairies on Oahu had heptachlor levels in their milk that were three to six times the acceptable state level. Two months after the contamination was detected, Hawaii's state health director ordered all fresh milk removed from stores and schools.[38]

During those two months, Hawaiians were exposed to high levels of the pesticide. Leland Parks, a scientist at the Pacific Biomedical Research Center in Honolulu, tested samples of breast milk from nursing mothers and found that the average level of heptachlor contamination had increased fourfold.[39] He said, "It appears that there is a relationship between exposure to the contaminated store-bought milk and the increase in human breast milk. . . . There is enough information to make me and many others uncomfortable."[40]

Another scientist suggested that the infants exposed to milk with high levels of heptachlor be followed to assess the long-term health risks of the exposure. According to the scientist, "A natural experiment has been foisted on the people of Hawaii."[41]

In 2016, a research team in Japan was ready to evaluate the results of this "natural experiment" to see whether there was a connection between heptachlor exposure and Parkinson's disease.[42] By coincidence, Japanese researchers had launched a Honolulu Heart Program in the 1960s to follow more than 8,000 men of Japanese ancestry living on Oahu for the development of heart disease. As part of that study, individuals completed diet surveys including questions on milk consumption. A subset had agreed to have autopsies performed at the time of their death. The research team examined the brains of 449 of them.

The investigators' findings were remarkable. They found that the density of nerve cells in the substantia nigra, the area of the brain affected by Parkinson's, was lowest in those who consumed the most milk.[43] Researchers also found a potential clue. The brains of those with Parkinson's were more likely to have the remains of heptachlor.[44]

The study had limitations. One was that the researchers did not have samples of milk that the participants may have drunk during the heptachlor contamination.[45] But its findings were supported by other studies. Some have found that levels of a related pesticide called dieldrin—which was widely sprayed on corn and cotton in the United States until 1970 (all uses were banned in 1987)—are also more likely to be found in the brains of people with Parkinson's.[46] The blood levels of these pesticides that dissolve in fat, such as DDT, heptachlor, and dieldrin, are also higher in individuals with Parkinson's.[47] Finally, in animal experiments, dieldrin kills dopamine-producing nerve cells.[48]

These pesticides have saturated the globe. While DDT, heptachlor, and dieldrin were banned from use on crops in the United States and other industrialized countries in the late twentieth century, use of the pesticides shifted to less industrialized nations, including India and China.[49] Even though China, for example, has now banned these pesticides, it used to be a major producer and consumer of them. As a result, the Chinese have high concentrations of the residues in milk from nursing mothers and increasing rates of Parkinson's.[50]

The effects may be felt for years to come, especially in middle- and low-income nations. Remnants of the chemicals are still found in the milk supply of many countries, including Brazil, China, Ethiopia, and Uganda.[51]

The accumulation of these pesticides in the body is not limited to adults who were exposed. They can be passed on to subsequent generations. As the Hawaii case showed, nursing women who ingest these chemicals from dairy or meat products can deliver them to their babies through their breast milk (**Figure 2**). DDT and similar pesticides were found in the breast milk of women living in Spain, Nicaragua, Taiwan, and the Spanish Canary Islands as recently as 2014.[52]

This class of pesticides may also cross the placenta into the developing fetus. Dieldrin then accumulates in body fats and is detectable in the brains of fetuses.[53] Twenty years ago, a report said that

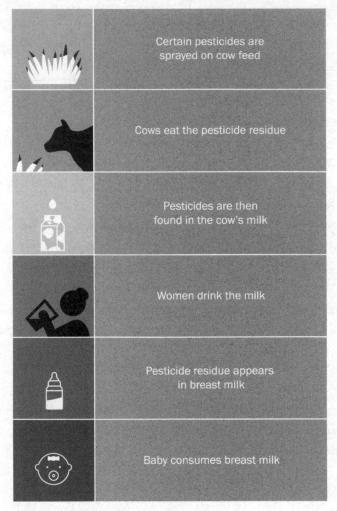

FIGURE 2. How pesticides can be passed from cow feed to babies.[54]

this "could pose a risk to the developing brain."[55] Such exposure may impair "motor and cognitive development in newborns and infants."[56] The long-term effects of DDT, heptachlor, and dieldrin on fetuses and breastfed children—and whether they can increase risk of developing Parkinson's—are unknown but concerning.

The United States banned DDT, Agent Orange, and heptachlor in the 1970s and 1980s. However, the United States has not banned all pesticides linked to Parkinson's. The one with perhaps the strongest link to the disease is still in widespread use: paraquat.[57]

THE DANGEROUS PESTICIDE COATING OUR CROPS TODAY

Paraquat has been used as a pesticide since the 1950s and is marketed as an alternative to the world's most popular weed killer, glyphosate, more commonly known as Roundup.[58] Paraquat takes care of weeds that not even Roundup can kill.[59] Today, it is used on farm fields across the United States (**Figure 3**), and its use continues to rise.[60] According to the US Geological Survey, its primary uses are for corn, soybeans, wheat, cotton, and grapes.[61]

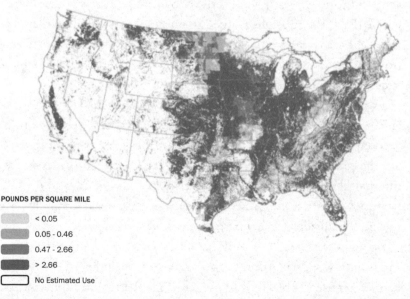

POUNDS PER SQUARE MILE

- < 0.05
- 0.05 - 0.46
- 0.47 - 2.66
- > 2.66
- No Estimated Use

FIGURE 3. Preliminary estimated agricultural use of paraquat in the United States, 2016.[62] Map created by the US Geological Survey. High estimate depicted in map includes more extensive estimate of pesticide use not reported in surveys.[63]

Paraquat may be an excellent weed killer, but its effectiveness comes at an enormous price. A 2009 study found that exposure to paraquat and another pesticide called maneb within five hundred meters of one's home increased the risk of Parkinson's disease by a whopping 75%.[64] Two years later, another study found that people who used paraquat—most notably farmers—were 2.5 times more likely to have Parkinson's than those who did not.[65] A scientist at the National Institutes of Health who has investigated paraquat said that the data were "about as persuasive as these things can get."[66]

In the laboratory, paraquat reproduces the features of Parkinson's disease.[67] In 1999, scientists at the University of Rochester gave paraquat to mice, and their activity decreased. Paraquat also killed dopamine-producing nerve cells in the rodents' substantia nigras, the same area of the brain affected by Parkinson's in humans. The greater the amount of paraquat administered, the greater the number of nerve cells lost. The effects mirrored those observed when the researchers gave MPTP, the toxin first linked to Parkinson's a decade earlier. The researchers concluded that the widely used herbicide causes destruction of dopamine-producing nerve cells in the substantia nigra and a syndrome similar to that caused by MPTP.[68]

Even beyond Parkinson's, paraquat has substantial safety concerns. A 2011 report called it "the most highly acutely toxic herbicide to be marketed over the last 60 years."[69] If it touches your eyes, it can damage the cornea and lead to blindness.[70] If you breathe it in, it can cause internal bleeding.[71] If you swallow a teaspoonful, it is fatal. In fact, paraquat is so toxic and so readily available that it is commonly used to commit suicide in many parts of the world.[72]

Because of its link to Parkinson's and its role in the deaths of thousands worldwide, thirty-two countries, including China, have banned its use (**Table 1**).[73] However, while Britain prohibits the use of paraquat, a company there continues to export it to the rest of the world.[74] The company indicates that the evidence of whether paraquat increases the risk of Parkinson's is "fragmentary and insufficient."[75]

COUNTRY	YEAR
Sweden	1983
Kuwait	1985
Finland	1986
Germany	1991
Slovenia	1997
Cambodia	2003
Ivory Coast	2004
Syria	2005
United Arab Emirates	2005

TABLE 1. List of when select countries banned paraquat[76]

Colombia, Ecuador, Guatemala, India, Indonesia, Japan, Mexico, Panama, Singapore, South Africa, Taiwan, and the United States are all customers.[77]

Advocacy groups within and outside the Parkinson's community have sought to ban paraquat in the United States.[78] On July 24, 2017, the Unified Parkinson's Advocacy Council, which includes the American Parkinson Disease Association, the Davis Phinney Foundation, The Michael J. Fox Foundation, and the Parkinson's Foundation, wrote to the EPA, "We write to express our concern with paraquat dichloride, which is shown to increase the risk of Parkinson's disease. We ask the Environmental Protection Agency to deny reregistration of this herbicide based on strong evidence of paraquat's harm to human health."[79]

For pesticides to be sold or distributed in the United States, they have to be registered with the EPA. According to the agency's own rules, registration is contingent on "scientific studies showing that they can be used without posing unreasonable risks to people or the environment."[80]

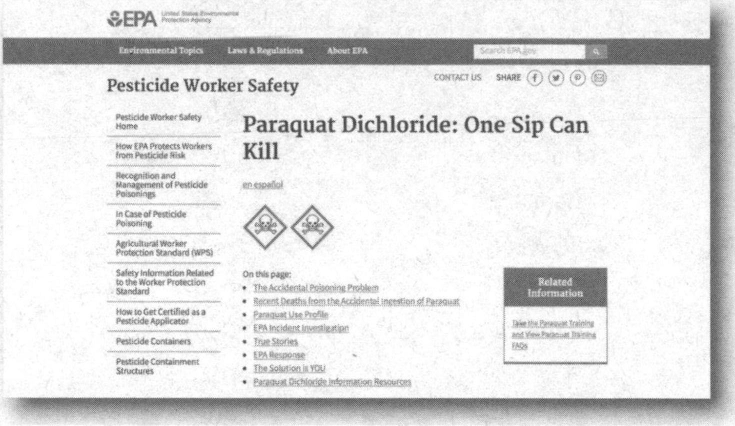

FIGURE 4. The US Environmental Protection Agency's warning about paraquat on its website.[81]

The EPA itself acknowledges the "high toxicity of paraquat."[82] Its own website reads, "Paraquat Dichloride: One Sip Can Kill" (**Figure 4**).[83] It includes "true stories" involving deaths from accidental ingestion of the chemical, including that of an eight-year-old boy who in 2008 drank paraquat that had been put in a Dr. Pepper bottle that he found in the garage. The child died in the hospital sixteen days later.[84]

The EPA reviews the safety standards of all herbicides like paraquat every fifteen years. In 2017, the EPA opened paraquat for reconsideration, and it has until October 2022 to make a decision on its future.[85] In 2019, in the absence of action by the EPA, The Michael J. Fox Foundation submitted a petition with over 100,000 signatures to the EPA urging it to ban paraquat.[86]

Meanwhile, paraquat use in the United States has doubled over the last decade (**Figure 5**).[87] It is now, according the EPA, "one of the most widely used herbicides registered in the United States."[88] In 2016, more than 12 million pounds of it were sprayed in the United States.

Appropriate equipment can minimize the harmful health effects of pesticides on humans. For example, protective gloves can reduce the risk of the disease among farmers who are exposed to some (including

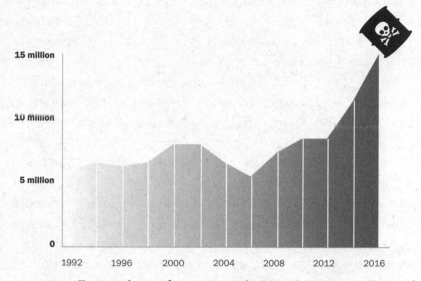

FIGURE 5. Estimated use of paraquat in the United States in millions of pounds, 1992–2016.[89]

paraquat), but not all, pesticides.[90] This kind of equipment and additional protective measures, such as boots, disposable coveralls, and goggles, could decrease the risk of developing Parkinson's disease in the United States and globally.

CASE FOR CAUSATION

In 1964, a decade after demonstrating that smoking causes lung cancer, Sir Austin Bradford Hill gave a lecture titled "The Environment and Disease: Association or Causation."[91] In it, the world's leading medical statistician outlined the criteria that an observed association—those who are exposed to pesticides, for example, have a higher rate of disease than the general population—would have to meet to demonstrate causation.[92]

Generally, an association—people exposed to certain chemicals have higher rates of Parkinson's, for example—does not prove that one causes the other. This was the tobacco industry's argument: that just because smokers had higher rates of lung cancer was not proof that

smoking was the cause.[93] Hill's achievement was to show that population studies could prove that one led to the other. The link between
the two had to be strong, it had to be supported by other scientific
evidence, and it had to meet other criteria.

Chemical companies have pushed back against the science. Following the tobacco industry playbook, according to a 2018 investigation
by the *New York Times*, they have lobbied the EPA not to ban certain pesticides by undermining population studies that show the link
to various neurological problems, including Parkinson's.[94] Pesticides
and other environmental factors tick most, if not all, of Hill's criteria.
The evidence includes numerous studies demonstrating a strong link
between certain pesticides and Parkinson's, high rates of the disease
among those with the greatest exposure (e.g., farmers), and animal
studies that replicate the features of the disease in exposed mice.[95]
Based on his criteria, we can conclude that certain pesticides are not
merely associated with Parkinson's. They likely cause it.

Not surprisingly, this conclusion is not shared by those who sell
pesticides. A 2016 review funded by a chemical company that makes
pesticides reached a different conclusion with regard to satisfaction of
these criteria and the link to causation.[96] That review concluded, "There
may be risk factors associated with rural living, farming, pesticide use
or well-water consumption that are causally related to [Parkinson's
disease], but the studies to date have not identified such factors."[97]

The solution to helping end Parkinson's disease is not to expose more
people to chemicals that are linked to the disease and produce the features of Parkinson's in the lab. The solution is to stop using such chemicals. Other countries with lower incomes and less strict environmental
policies have banned paraquat. The United States should do the same.

ECHOES OF *SILENT SPRING*

Rachel Carson launched the modern environmental movement in
the 1960s with the publication of *Silent Spring*.[98] In the book, she described the impact of the indiscriminate use of pesticides. She wrote,

It is ironic to think that man might determine his own future by something so seemingly trivial as the choice of an insect spray.

All this has been risked—for what? Future historians may well be amazed by our distorted sense of proportion. How could intelligent beings seek to control a few unwanted species by a method that contaminated the entire environment and brought the threat of disease and death even to their own kind? Yet this is precisely what we have done.[99]

Carson went on, "The public must decide whether it wishes to continue on the present road, and it can do so only when in full possession of the facts."[100] Sixty years ago, facts were in short supply, and many consequences were unknown.

Now we know. We recognize the benefits of pesticides in agriculture, but we also know that certain kinds of these chemicals are linked to Parkinson's, not to mention autism, lung disease, and various cancers.[101]

When there are other, less toxic options, we have no excuse for using chemicals that make us sick. We can no longer claim ignorance. If we continue to allow the use of these pesticides, we are choosing to permit the spread of Parkinson's through the farms we tend, the air we breathe, and the water we drink.[102]

CLEANING UP
How Solvents and Contaminated
Groundwater Spread the Disease

The benefit of the doubt should go to the people, not
the chemical.

—Retired Master Sergeant Jerry Ensminger[1]

IN 1988, DANNY FROMM WAS A TYPICAL SOUTHERN CALI-
fornian teenager who enjoyed working on cars, especially his red
1972 Chevy Nova with a black top. He paid $1,300 for it, money
he had saved from working as a gas station attendant at the local
Unocal 76. Fromm and his buddies replaced the car's engine, which
made it "really awesome and really fast," he recalled. The car gave the
seventeen-year-old the freedom that he craved. Freedom to do what
he wanted, when he wanted. Freedom that he no longer has.

Straight out of high school, Fromm began working in the aerospace
industry, cleaning circuit boards with the solvent trichloroethylene
(TCE). Never warned of any risk or provided with protective gear,
he inhaled the sweet-smelling chemical and exposed his skin to it for
eight hours a day over the course of a decade.

When Fromm was thirty-five, he noticed that his right pinky kept
twitching. "Stress," his first doctor said and recommended a little

wine. His second doctor gave him a much different diagnosis: Parkinson's disease.

After Fromm started levodopa, his symptoms did improve. But after five years, he developed involuntary movements of his head, neck, and trunk, a common side effect of the medication. To control these movements and improve his Parkinson's, he underwent a surgery called deep brain stimulation. As the name suggests, wire electrodes were inserted into his brain and connected to a battery-powered stimulator positioned just under the skin in his chest. The surgery alleviated some of his symptoms, including his tremor, but it was far from a cure.

Fromm is now forty-eight and lives in Idaho with his wife and six-year-old son, Logan. He also has two older sons who live nearby with their mother. In the morning, Fromm has trouble walking. He has what he calls a "hard-core shuffle." So he takes his medication immediately after he wakes up, and he does leg exercises to relieve his leg stiffness. He is then able to get out of bed, shower, and dress without assistance.

Each morning, he makes Logan's lunch and takes him to school. By the time Logan comes home, Fromm's shuffling has usually returned. Sometimes he feels embarrassed for himself and for his son. Some days are better than others, though. On a good day, he can play with Logan, mow the lawn, and even walk on trails near his home. But Fromm says the effects of his medications are hit-or-miss. Many days his mobility is severely limited.

He and his son spend a lot of time together. Occasionally, Logan will mimic Fromm's walk, even his shuffle, in a good-natured way. His son rarely speaks of the disease, though. Fromm says, "I better have a good relationship with Logan. He is going to take care of me when I am older."

Fromm regrets every day that he stayed at his aerospace job, working with trichloroethylene (sometimes called "tri" or "trike"). While the solvent has not been proven definitively to cause Parkinson's, those who are exposed to it at work are six times more likely to develop the disease than those who are not.[2] "If you're working with it," Fromm says, "quit your damn job and get away from it."

But you do not have to have had a job cleaning circuit boards to have been exposed to the dangerous chemical. Almost all of us have been.

Trichloroethylene was introduced as a chemical in the 1920s. It soon found many commercial and consumer applications, from flushing rocket engines to cleaning carpets.[3] Because TCE readily evaporates and can be inhaled, it was also used as an anesthetic in surgery and childbirth.[4] Due to its toxicity, the Food and Drug Administration (FDA) banned the use of TCE as an anesthetic in 1977.[5] In the 1990s, the first published reports linked the chemical to Parkinson's.[6]

Today, TCE is still used to degrease metal and as a spot-cleaning agent in dry cleaning. It is also an ingredient in many common household products, including paint removers, glue, stain removers, carpet cleaners, and gun cleaners.[7] The estimated annual use of TCE in the United States is 250 million pounds.[8]

WIDESPREAD EXPOSURE

Though Fromm was allowed to work without any protection, TCE's harms were already known in industry.[9] As far back as 1932, Dr. Carey McCord, a medical advisor for Chrysler Corporation in Cincinnati, Ohio, wrote a letter to the *Journal of the American Medical Association*. It was titled the "Toxicity of Trichloroethylene" (**Figure 1**).[10]

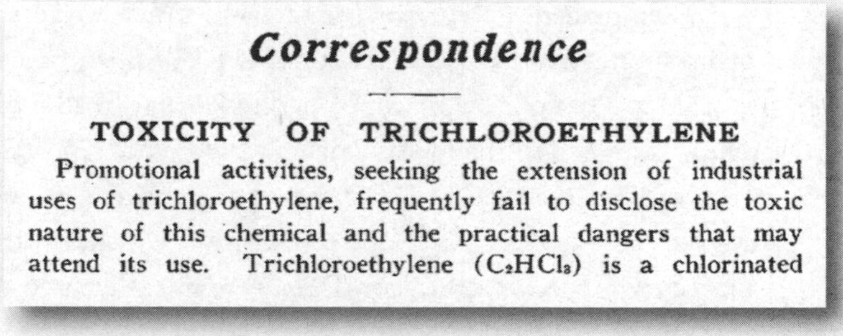

Correspondence

TOXICITY OF TRICHLOROETHYLENE

Promotional activities, seeking the extension of industrial uses of trichloroethylene, frequently fail to disclose the toxic nature of this chemical and the practical dangers that may attend its use. Trichloroethylene (C_2HCl_3) is a chlorinated

FIGURE 1. Dr. Carey McCord's 1932 letter to the *Journal of the American Medical Association* warning of the dangers of the solvent.

McCord began his letter, "Promotional activities, seeking the extension of industrial uses of trichloroethylene, frequently fail to disclose the toxic nature of this chemical and the practical dangers that may attend its use." He went on, "In industry, trichloroethylene may enter the body through the breathing of vapors or through the skin." McCord detailed the lethal effects of different concentrations of TCE from inhalation and skin absorption, based on studies done on rabbits. He concluded, "Any manufacturer contemplating the use of trichloroethylene may find in it many desirable qualities. Too, in the absence of closed systems of operation, he may find in this solvent the source of disaster for exposed workmen."[11]

Decades later, that disaster arrived at a small plant in Berea, Kentucky. Workers there dipped their arms into a vat of TCE to clean small metal parts without wearing any protective gear. Two men who did this work, breathing in the solvent's vapors over the course of twenty-five years, developed Parkinson's. A third worker, who sat at a workstation adjacent to the TCE vat, was also diagnosed with Parkinson's. With her bare hands, she received parts that the first two workers had cleaned with TCE.[12]

At a doctor's appointment, one of the men mentioned that his co-workers had also developed Parkinson's.[13] Because of existing concerns about the toxic effects of TCE on the nervous system, the doctor and her colleagues at the University of Kentucky decided to investigate.[14] They mailed a questionnaire to 134 former workers of the plant; 65 responded. Of these, an astounding 68%—forty-four of them—reported at least one symptom of parkinsonism. In addition, the researchers examined thirteen respondents who did not report any symptoms and found that the speed of their hand movements was "significantly slower" than normal.[15] The investigators concluded that those employees who worked closest to TCE were most likely to develop signs of the disease.

The researchers then decided to see if they could replicate what happened to the plant's employees in laboratory animals. They fed TCE to rats for six weeks, and the effect was dramatic. The animals lost almost half of their dopamine-producing nerve cells in the substantia nigras of their brains.[16]

And yet, from 1981 to 1991, the production of TCE increased *one thousand fold*.[17] It was nearly ubiquitous in American industry in the second half of the twentieth century.[18] A National Institutes of Health–funded epidemiological study that linked work exposure to TCE to Parkinson's concluded that the "potential public health implications are substantial."[19]

While US production has decreased from its peak, millions of pounds of TCE continue to be released into the environment annually. It can be found in the air, the soil, food, and human breast milk.[20]

COVERING UP A CATASTROPHE

Occupational exposure to TCE extends to those in the military. Perhaps the worst example occurred at the Marine Corps Base Camp Lejeune in Jacksonville, North Carolina. Since 1941, the base, named for a World War I marine general, has trained and maintained combat-ready marines to be "the world's best war fighters."[21] It currently has a population of 170,000, including those on active duty, retirees, dependents, and civilians.

For over two decades, from 1953 to 1987, the residents of Camp Lejeune drank and bathed in toxic water.[22] During that time, more than 70 chemicals, including TCE and a similar solvent used by dry cleaners called perchloroethylene (PCE)—also linked to Parkinson's—poisoned the base and its water supply.[23] The marine base needed clean metal parts for all its tanks, airplanes, and amphibious vehicles, and TCE was the answer.[24]

Officers also needed spotless uniforms. ABC One-Hour Cleaners, located near the base, cleaned many of them. It also, according to the Environmental Protection Agency (EPA), "improperly disposed" of its waste.[25] The dry cleaner was not alone in its sloppy practices. On-base spills and leaks from underground storage tanks also contributed to the mess.[26] The result was the discovery of approximately one ton of waste in the soil and groundwater at the base.[27] The concentration of chemicals, including TCE and PCE, in the drinking water was 240 to 3,400 times the level permitted by safety standards.[28]

From 1980 to 1984, the Marine Corps leadership received numerous warnings about the water's contamination.[29] Despite the multiple notices, it failed to act.[30] The contaminated wells on the base remained open, exposing residents every time they drank water, bathed, swam, cooked, or cleaned. In 2010 a US House of Representatives Oversight Subcommittee found, "For thirty years, Marines and their dependents serving at Camp LeJeune were exposed to toxic chemicals in their drinking water. It took the [US Marine Corps] more than four years to shut down drinking water wells they knew to be contaminated with toxic chemicals and another 24 years and an act of Congress to force them to inform veterans about this contamination of potential health problems. For two decades the U.S. Marine Corps prevented full disclosure regarding the true extent of contamination at Camp Lejeune."[31] As many as 1 million individuals ended up coming into contact with the toxins on the base.[32] According to a 2005 National Academy of Sciences panel, it was "the largest human exposure to [trichloroethylene] from drinking water in this nation's history."[33]

Because TCE readily evaporates into air, the drinking water was not the only hazard. Vapors from the solvent can migrate through the soil and into nearby buildings, affecting the indoor air quality.[34] Some of the Camp Lejeune barracks were found to be polluted.[35]

Lori Lou Freshwater, an investigative journalist, has written extensively about the contamination of Camp Lejeune.[36] Her family was also one of its victims. As a child, Freshwater lived with her family on the base from 1980 to 1983 during the peak contamination. There, she wrote, "my entire childhood was consumed by tragedy. The chemical contamination can be linked to the deaths of my two baby brothers . . . and to my mom's own difficult final years, when she was dying from two types of leukemia."[37] Today, Freshwater says, "a lot of people have no awareness that Parkinson's disease can come from these environmental exposures."

The denials and delays among the upper echelons of the Marine Corps only increased the health risks to the Lejeune population. A local cemetery is filled with scores of babies, infants, and children who died due to the contamination.[38] According to one news report, "Hundreds of mothers suffered miscarriages or gave birth to stillborn babies or infants with birth defects. . . . An unknown number, but likely thousands, have developed cancers . . . and Parkinson's disease after living on the base."[39]

In 2017, the Department of Veterans Affairs (VA) added Parkinson's to the conditions considered to be "presumptively" due to time spent at Camp Lejeune.[40] The VA said that its "review resulted in the recognition that liver cancer and Parkinson's disease . . . are conditions for which there is strong evidence of a causal relationship and evidence that the condition may be caused by exposure to contaminants."[41]

TOXIC VALLEY

Silicon Valley is home to Google, LinkedIn, Yahoo, and other technology titans. It is also home to more EPA Superfund sites than any place in the country (**Figure 2**).[42]

In the 1960s and 1970s, Silicon Valley's legendary semiconductor companies, including Fairchild Semiconductor and Intel, used TCE to clean silicon chips.[43] The soil and groundwater are now tainted with the chemical, which evaporates into the air.[44]

This contamination was typical of the era. In the 1970s, thousands of toxic waste dumps were polluting the air, land, and water of cities and towns throughout the country. In response, Congress established the Comprehensive Environmental Response, Compensation, and Liability Act in 1980, which created a cleanup program that became known as "Superfund."[45] The act allowed the EPA to designate sites for cleanup and forced the parties—often companies—responsible for the contamination to either do the work themselves or reimburse the government for the cost of doing so.[46]

SUPERFUND SITES IN SILICON VALLEY

Here are 15 of the 21 Superfund sites in Silicon Valley that are contaminated with TCE

1 CTS Printex, Inc
Plymouth & Colony STS
Mountain View, CA 94043

2 Teledyne Semiconductor
1300 Terra Bella Ave
Mountain View, CA 94043

3 Spectra-Physics
1250 W Middlefield Rd
Mountain View, CA 94042

4 Moffett Field Naval Air Station
Moffett Field, CA 94035

5 Fairchild Semiconductor Corp.
369, 515 N Whisman,
313 Fairchild & 401 National
Mountain View, CA 94942

6 Raytheon Corp.
350 Ellis St
Mountain View, CA 94943

7 Westinghouse Electric Corp.
401 Hendy Avenue
Sunnyvale, CA 95117

8 TRW Microwave, Inc. (Building 825)
825 Stewart Dr
Sunnyvale, CA 94086

9 Advanced Micro Devices, Inc.
901 Thompson Place
Sunnyvale, CA 94086

10 Monolithic Memories
1165 E Arques Ave
Sunnyvale, CA 94086

11 National Semiconductor Corp.
2900 Semiconductor Ave
Santa Clara, CA 95050

12 Intel Magnetics
3000 Oakmead Village Dr
Santa Clara, CA 95051

13 Applied Materials
3050 Bowers Ave
Santa Clara, CA 95051

14 Synertek, Inc. (Building 1)
3050 Coronado Blvd
Santa Clara, CA 95051

15 Intel Corp. (Santa Clara III)
2880 Northwestern Pkwy
Santa Clara, CA 95051

FIGURE 2. Map of fifteen Superfund sites in Silicon Valley contaminated with TCE.[47]

In the Silicon Valley town of Mountain View, Google sits right on top of one Superfund site. The cleanup of the area, which was polluted by previous occupants, is currently incomplete and will continue for many decades.[48] In the meantime, residents, who own some of the most expensive real estate in the country, risk being exposed to TCE.

When one resident, Jane Horton, bought her house near Google's current location, she had no idea what was in the ground.[49] After learning that she lived across the street from a Superfund site, she had her indoor air tested and found that it was contaminated, as she wrote in a 2014 piece in *The Guardian*. She also discovered that a toxic plume of TCE ran right under her home.[50] Even after 75% of the TCE was removed from the groundwater at the Superfund site, the chemical's levels in Horton's indoor air were still above the EPA's safety threshold. Consequently, her home had to be "cut, butchered, vented and fanned with tubes and vents."[51]

FIGURE 3. Jane Horton outside her home in Mountain View, California, with her TCE-remediation system, 2010. *Courtesy of Michelle Le/*Mountain View Voice.

In 2002, a local newspaper, the *Mountain View Voice*, reported that six residents who lived on or around Walker Drive, a short street in Horton's neighborhood, had been diagnosed with Parkinson's disease.[52] In addition, four others who lived nearby had brain tumors. (The EPA classifies TCE as a carcinogen.)[53]

Today, Horton lives in her same home, with her air-remediation system running twelve hours a day (**Figure 3**). It funnels TCE-contaminated air from under her house to above her house. She has her air tested through the EPA once or twice a year and says that her indoor "air is probably the cleanest in Mountain View."

While she has little concern about her own air, Horton worries about the community. She says, "Who is benefiting from this contamination? Why can't we just clean this up? . . . [It is] so frustrating that people's lives are put at risk."

Tainting up to 30% of US drinking water supplies, TCE is now the most common organic (i.e., carbon-containing) contaminant in groundwater in the United States.[54] As of July 3, 2018, the EPA had 1,346 sites on its National Priorities List eligible for remediation and financing under the federal Superfund program.[55] Of these, almost half are contaminated with TCE.[56] The map in **Figure 4** and the website www.endingPD.org show the locations of TCE-contaminated Superfund sites throughout the United States.

A LOCAL STORY

Superfund sites are only the beginning. Thousands of additional locations across the country are contaminated with the chemical. "Superfund sites are [only] a fraction" of the areas contaminated with TCE, according to Lenny Siegel, the former mayor of Mountain View and the executive director of the Center for Public Environmental Oversight.

The TCE leaching through the ground in Victor, New York, which is just fifteen minutes from the home of one of our authors, Ray Dorsey, is one example.[57] In 1990, the New York State Department of Health began sampling small community water supplies across the

FIGURE 4. Map of active US Superfund sites contaminated with TCE, 2018.

state, including in Victor, which has a population of 14,000. The water supply for many of the residents came from a local spring that was found to be contaminated with TCE at levels more than twice what is considered safe.

The contamination may have begun ten years earlier and was traced to a local sand and gravel pit. Inspectors found a mile-long TCE plume emanating from the pit that has contaminated a handful of local private wells, increased TCE levels in the soil, and, in six nearby homes, raised the TCE levels in the air above safe levels.[58]

Jamie Myers (not her real name) is a fifty-one-year-old nurse who lives in Victor. Her house was supplied by the contaminated spring, and for ten years, four times a week, she used to jog up the steep incline to the TCE-laced gravel pit. In 2014, she noticed that she could no longer make it up the hill. She was later diagnosed with Parkinson's.

Myers became aware of the history of TCE contamination in 2008, and she always suspected that her Parkinson's might be due to environmental factors. She has no family history of the disease, eats a salad for lunch, and has been a lifelong runner. Today, she says, "I am frustrated because [the contamination] has become personal to me."

Now that her home has an air-remediation system that moves TCE from under her house to a vent above, she is less worried about her three kids. Her water, like everyone else's in Victor, now comes from a municipal source instead of a well. She says, "If I were worried about my kids, we'd get out right away."

According to local reports, a 2009 cancer study in the area was "inconclusive" but did show an "unusual number of brain tumors." The pattern is similar to what was found in Jane Horton's Mountain View neighborhood.[59] The cancer study did not asses the risk of Parkinson's.

TCE contamination is not just a US problem. Global consumption of the chemical is projected to increase 2% annually. In China, where Parkinson's rates are increasing most rapidly, use of the solvent is projected to increase 4% per year.[60]

Parts of Europe have banned TCE altogether.[61] In late 2016 and early 2017, the EPA proposed banning its use for spot-cleaning in dry cleaning facilities and for commercial degreasing, such as cleaning motors. The agency had several health concerns, including the fact that TCE is "carcinogenic to humans by all routes of exposure" and that chronic exposure is harmful to the brain and nervous system.[62]

Industry groups sought to delay or prevent regulatory action.[63] The National Cleaners Association, a dry cleaning trade group, has argued that alternatives to TCE are not as effective.[64] The Association of Global Automakers said that the EPA did not follow sound economic principles in its analysis.[65] Then, in late 2017, the EPA announced that it would postpone its proposed ban indefinitely.[66]

Those who have been and are being exposed to TCE are tired of waiting. Senator Tom Udall, whose uncle, Congressman Morris Udall, had Parkinson's, is pressuring the EPA to ban TCE. In August 2018, he held a press conference with families. They shared stories of loved ones passing away from diseases that have been linked to the solvent and emphasized the need for swift action. EPA administrator Andrew Wheeler said, "Absolutely, we need to be moving forward to do something on TCE and these other chemicals."[67] As of July 2019, the EPA had yet to act.

HOPE FOR ENDING PARKINSON'S

Toward the end of his lecture on causation, Sir Austin Bradford Hill said, "In occupational medicine our object is usually to take action."[68] These actions can improve health.

The Netherlands is one of the few countries in the world where rates of Parkinson's disease are actually waning.[69] A 2016 study found that from 1990 to 2011, the number of new cases of the disease "decreased sharply."[70]

The reason is not known for sure, but just as the skyrocketing rates of Parkinson's tracked with industrialization, the disease's auspicious fall follows the country's efforts to clean up. The Netherlands, along with other European countries, was among the first to ban paraquat.[71] Use of other pesticides linked to Parkinson's decreased or stopped altogether. As a result, levels of DDT, dieldrin, and their metabolites in the fat, blood, and breast milk of Dutch citizens also went down. For example, from 1968 to 1986, fat levels of dieldrin decreased by about 75% and levels of DDT by 90%.[72] The Netherlands also banned TCE, and in 1981 the TCE levels in the air were among the lowest in all of Europe.[73] Air pollution in general—also linked to Parkinson's—has also decreased substantially.[74] From 1990 to 2012, emissions of multiple air pollutants decreased by 50% or more.[75]

Many degenerative and man-made diseases reflect the environment that our parents and grandparents created. Just as environmental contamination can worsen health and create disease, cleaning up the contamination can have the reverse effect. Enlightened by science, we now can rectify any prior shortcomings for ourselves and our descendants. But we, as Hill suggests, must act.

PROTECTING OURSELVES
The Role of Head Trauma, Exercise, and Diet

> Muhammad is battling a relentless, remorseless, insidious thief. Parkinson's recognizes no title, respects no achievements, nor bows to any amount of talent, courage, or character. Parkinson's does not discriminate. There is no question that Parkinson's is the fight of Muhammad's life.
>
> —Lonnie Ali, in testimony to Congress in 2002[1]

FOR MOST OF HIS LIFE, MUHAMMAD ALI RAISED OUR CONsciousness. He did so about the Vietnam War, racism, identity, religious freedom, and, finally, the health risks of head trauma.

On December 11, 1981, Ali took on Trevor Berbick in his sixty-first and final professional fight. Three years later, he was diagnosed with Parkinson's. For twenty-seven years, Muhammad Ali fought boxers. For the next thirty-two, he would be in the ring with Parkinson's disease. At the time of his diagnosis, little was known about the link between head trauma and developing the disease.[2] Since then, multiple studies have found that repetitive blows to the head increase the risk.[3]

In 2006, researchers discovered that even a single head injury resulting in loss of consciousness or amnesia tripled the risk of Parkinson's.[4] Repeated head trauma raised the risk even further.[5] Traumatic brain injury increases not only the chances of developing the disease but also the rate of progression of parkinsonian features. It also leads to the accumulation of Lewy bodies, those clumps of misfolded proteins found in the brains of people with Parkinson's.[6] And just as the combination of particular gene mutations and chemical exposure can up our odds, so too can the risk of traumatic brain injury be amplified by environmental pressures. Head trauma plus exposure to the pesticide paraquat almost triples a person's risk of the disease.[7]

Football players, especially professional ones, are, of course, vulnerable. They are more likely to develop neurodegenerative disorders, including Alzheimer's disease, amyotrophic lateral sclerosis (ALS), and chronic traumatic encephalopathy (CTE), which, resulting from repeated brain trauma, can lead to dementia.[8] The link between playing professional football and Parkinson's disease, however, is less clear. A 2012 study found that National Football League (NFL) players face an almost three times greater risk of death due to conditions like ALS—Lou Gehrig's disease—and Alzheimer's but not Parkinson's.[9]

But a 2017 study of football players, published in the *Journal of the American Medical Association*, found a connection. The risk of CTE increased with the duration and intensity of a person's football experience and was present in the brains of almost every former NFL player in the study—110 out of 111. It also showed up in people who had played high school football, but in fewer of them—three out of fourteen.[10] And among the former professional players with CTE, two-thirds had parkinsonian features, including slowness, a tendency to fall, and tremors (**Figure 1**). Six of the 111 had been diagnosed with Parkinson's.[11]

Fueled by these concerns, some NFL players are making big life changes. Chris Borland, a rookie linebacker who led the San Francisco 49ers in tackles in 2014, is one example. At the age of twenty-four, he walked away from the remainder of a nearly $3 million contract and the life of a professional football player.[12] In the *Huffington Post*, the former University of Wisconsin all-American wrote, "Had I stopped playing football following high school, I wouldn't understand why a

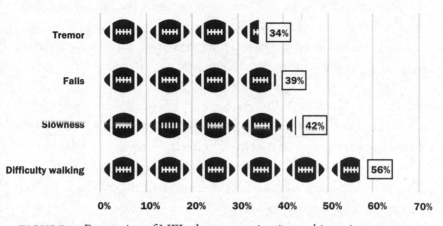

Figure 1. Proportion of NFL players experiencing parkinsonian symptoms in a 2017 study.[13]

24-year-old would quit the NFL just one year into his professional career. If I hadn't made hundreds upon hundreds of tackles during college and my time in the pros, I might be susceptible to the thinking [that] a player can do so safely. My post–high school experiences gave the mounting and damning evidence that football causes brain damage far more gravity. The tragic stories I'd heard murmurings about since my high school days weren't outliers like I'd been led to believe."[14]

He was right. In 2009, after decades of denial, an NFL spokesman told the *New York Times*, "It's quite obvious from the medical research that's been done that concussions can lead to long-term problems."[15] In 2014, the NFL released documents in a court proceeding indicating that the organization expects nearly a third of its retired players to develop long-term cognitive problems at "notably younger ages" than the general population.[16] These admissions helped form the basis of a legal settlement in which the NFL agreed to provide $765 million in medical aid to more than 18,000 former NFL players.[17]

A year and a half after the settlement went into effect, claims for neurodegenerative disorders exceeded all expectations. According to a 2018 *Los Angeles Times* article, 113 retired players have already filed claims related to Parkinson's; 81 have been either paid or approved.[18] The number of claims far exceeds the projection that only fourteen claims for the disease would be paid over the sixty-five-year duration of the settlement.[19] In the first eighteen *months*, the number of Parkinson's claims was five times greater than the amount predicted for sixty-five *years*.

Over a career that began in 1956 and spanned fifteen seasons, Forrest Gregg played in 188 consecutive NFL games during his Hall of Fame career. He is just one of many former professional players who have been diagnosed with Parkinson's.[20] Gregg was a nine-time Pro Bowl offensive lineman for the Green Bay Packers and Dallas Cowboys who went on to coach the Cincinnati Bengals to their first Super Bowl appearance in 1982.

During his playing career, Gregg suffered countless concussions. In his sixties, Gregg began acting out his dreams at night, an early sign of Parkinson's. He dreamed that he was blocking for quarterback legend Bart Starr. Gregg ended up knocking his wife out of bed. In his seventies, Gregg developed a soft voice, tremors in his hands, and a stooped posture that led to his diagnosis. If he had known the risks, Gregg said that he would have still played football, though he may have shortened his career.[21] In 2019 Forrest Gregg died from complications of Parkinson's disease.[22]

Concussions are not just an issue for professional football players. According to one study, among youth football players aged eight to twelve, the rate of concussions is comparable to that reported for high school and college athletes.[23] Some studies report lower rates and some higher, especially for college athletes.[24]

Concussions in sports are generally underreported. For example, among 1,500 high school varsity football players in Wisconsin, 30% reported a previous history of at least one concussion, and 15% reported a concussion in that season. According to a confidential survey, less than half of the players reported their concussions that season to a coach, teammate, trainer, or parent.[25] Pressure from coaches, teammates, parents, or fans contribute to the silence. More than 25% of female and college athletes continue playing after a suspected concussion because of such pressure.[26]

Football, of course, is not the only sport with a high concussion rate. Boys' ice hockey and lacrosse and girls' soccer and lacrosse also have high rates.[27] However, football is the clear leader. In a national sample of US high school athletes for twenty sports, football accounted for 47% of all concussions.[28] Some efforts have been made to limit contact, enforce rules, and educate players.[29] These are likely not enough.

Military veterans also pay the price of traumatic brain injury. In a 2018 study, researchers examined the records of more than 300,000 individuals from a Veterans Health Administration database.[30] Nearly 1,500 individuals were diagnosed with Parkinson's after having a traumatic brain injury, often due to explosions. Mild traumatic brain injury in veterans increased the risk of developing Parkinson's by more than 50%.[31] More severe injury was associated with an even higher risk.

The burden of Parkinson's disease linked to traumatic brain injury is poised to grow. According to the US Department of Defense, nearly 400,000 service members have been diagnosed with a traumatic brain injury since 2000 (**Figure 2**).[32] Another 8 million veterans have likely

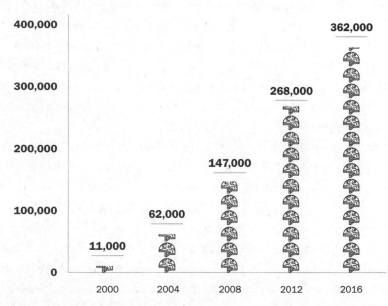

FIGURE 2. Cumulative number of US service members diagnosed with traumatic brain injury, 2000–2016.[33]

experienced such an injury.[34] For those with moderate or severe traumatic brain injury, one in fifty will likely develop Parkinson's within twelve years.[35] Traumatic brain injury adds to the risk of Parkinson's among veterans, many of whom have already been exposed to the pesticide Agent Orange and chemicals like TCE.

MOVE YOUR BODY

Published in 1899, the two-volume *Manual of Diseases of the Nervous System*, written by British neurologist Dr. William Gowers, used to be referred to as the "Bible of Neurology."[36] In it, Gowers recommended that for individuals with Parkinson's disease, "life should be quiet and regular, freed, as far as may be, from care and work."[37] Sir William Gowers never met Jimmy Choi.

Since 2012, nine years after his diagnosis with Parkinson's at the age of twenty-seven, Choi has completed one hundred half marathons, fifteen full marathons, one ultramarathon, six Gran Fondo bicycle rides, multiple Spartan Races, and countless 5K and 10K runs. When he was first diagnosed, Choi's life *was* quiet. He went into denial, hid his diagnosis even from his wife, and did nothing. By the time he was thirty-four, he weighed 240 pounds and walked with a cane. One day while carrying his young son, the two fell down a flight of stairs. Neither was hurt, but the accident motivated Choi to regain his health.[38]

After reading an article about someone with Parkinson's competing in a marathon, Choi took up running. He completed a 5K, then a 10K and a 15K, and, in 2012, his first marathon.

Several years ago, while watching the television show *American Ninja Warrior*, Choi's ten-year-old daughter told her father he should apply. Choi did, and in July 2017, he became the first person with Parkinson's to compete. Choi said participating was the "most terrifying thing [he has] done." Although he fell on the course's balance obstacle, his performance was stirring to watch. The show's co-host, Akbar Gbajabiamila, whose father has Parkinson's, was visibly

FIGURE 3. *American Ninja Warrior* competitor Jimmy Choi and co-host Akbar Gbajabiamila, a board member of The Michael J. Fox Foundation for Parkinson's Research.

moved (**Figure 3**). A year later, Choi returned to *American Ninja Warrior*. Despite a noticeable tremor, he competed until he was bested by the third of ten obstacles. The crowd gave him a standing ovation.[39]

Choi's exercises have a purpose. He runs to build endurance to prevent fatigue from Parkinson's. He does burpees so he can learn how to fall safely and get back up again. He only wishes that he had embraced exercise earlier.

Choi is not sentimental about his disease. He says, "Parkinson's sucks." But to people who suggest that he could run even faster without it, Choi replies, "I don't think that I would be running a marathon if I didn't have Parkinson's."[40]

He is not the only one with Parkinson's who has found exercise to be a boon. Cathy Frazier, a graphic design firm owner who was diagnosed in 1998 at the age of forty-three, has discovered the benefits of bicycling.[41] She did so with her friend, Dr. Jay Alberts, who happens to be a neuroscientist. Alberts, an avid cyclist, thought it would be fun to ride a tandem bicycle with Frazier 460 miles across the entire state of Iowa to raise awareness of her disease. During the ride, Alberts set a quick pace. "I was driving her to pedal faster than she would on her own."[42]

During their weeklong trip, both Frazier and Alberts noticed that her symptoms improved. Frazier said, "It doesn't feel like I have Parkinson's when I'm on the bike." Alberts saw that her tremor subsided and her handwriting improved. Given the changes, Frazier continues to cycle, sometimes on a tandem bicycle, and her writing remains smooth and clear.[43] The pair even co-founded Pedaling for Parkinson's, an organization dedicated to understanding how physical activity affects the motor symptoms of the disease.[44]

Alberts began to investigate what he termed "forced exercise"—exercising at rates higher than people can achieve by themselves.[45] In a small ten-person study, Alberts and his colleagues looked at individuals with Parkinson's who rode a tandem bicycle with a trainer pushing their speed, as he had done with Frazier. The scientists found that both forced exercise and traditional bicycling improved aerobic fitness. However, only forced exercise improved motor function in the study's participants.[46]

Alberts also looked into studies done on animals. In one study, mice with Parkinson's experienced a similar benefit from forced exercise. After the mice were put on a treadmill at a faster speed than their "preferred running velocity," their motor function improved, and growth factors for nerve cells increased.[47] These growth factors help nerve cells develop and survive. They may also increase the release of dopamine and improve communication between nerve cells.[48]

Alberts has continued his research. In a National Institutes of Health–funded, eight-week, one-hundred-person study, he found that people with Parkinson's who biked at high speed—between eighty and ninety revolutions per minute—appeared to have the greatest benefit from exercise. Next up is a one-year study that will follow three hundred people who are pushed beyond their usual exercise rates.

Alberts's positive results are supported by numerous randomized controlled trials that have demonstrated the value of exercise for people with Parkinson's—even at a pace they could achieve on their own. A review of fourteen randomized controlled trials of stretching, strength training, walking, and other exercises found that all improved physical function, quality of life, balance, and gait speed for people with the disease.[49]

Given the clear benefits of exercise for those with Parkinson's, a new question arises. Can exercise help prevent Parkinson's? Two studies published in 2018 sought to answer that question. In one, researchers examined over 7,000 veterans. They found that compared to the least physically fit, those who were in the best shape were 76% less likely to develop Parkinson's twelve years later. Their findings and others led the researchers to conclude, "[These] observations provide strong support for recommending physical activity to diminish [the] risk of Parkinson's disease."[50]

In the second study, researchers at Zhejiang University in China reviewed eight previous studies, involving over 500,000 participants, to determine whether physical activity lowered the chances of developing Parkinson's a decade or more later. They found that exercise was associated with a 21% decreased risk. Exercise that was moderate to vigorous and done consistently (e.g., 7 to 8 hours of walking or 3.5 to 4 hours of lap swimming each week) pushed that number even higher—to a 29% decreased risk of developing the disease.[51] An additional hour per week of vigorous exercise (e.g., jumping rope) or two hours of moderate exercise (e.g., bicycling) decreased the risk of Parkinson's by another 9%. Although promising, the protective effects of exercise are far from absolute. But science is showing that in addition to helping us feel better and live longer, exercise likely reduces our risk of Parkinson's.[52] And the more you do it, the better.[53]

EAT GOOD FOOD

The Mediterranean diet already has an excellent reputation. It reduces the risk of developing heart disease, memory loss, and cancer.[54] It also helps us live longer.[55] The diet centers on vegetables, legumes (e.g., peas, beans, lentils), fruits, cereals, and unsaturated fatty acids, especially olive oil. It emphasizes fish over dairy products, meat, and poultry, and it includes wine.[56] Dr. Alberto Ascherio, an Italian physician and epidemiologist at Harvard Medical School, and his colleagues wanted to know whether the diet's wonders extended to Parkinson's.

To find out, they looked at what 130,000 health professionals ate. The researchers characterized their diets as either "prudent" (high in fruit, vegetables, legumes, whole grains, and fish) or "Western" (high in red meat, processed food, refined grains, French fries, sweets, and high-fat dairy products). After sixteen years, there were 22% fewer cases of Parkinson's among the "prudent," or Mediterranean, diet group compared to those who ate typical American fare.[57] Subsequent studies have supported these findings.[58]

How exactly the Mediterranean diet may protect us is largely unknown.[59] One possibility is that antioxidants, which are vitamins and other substances that are abundant in the diet, may reduce the clumping of alpha-synuclein into Lewy bodies inside nerve cells.[60] Antioxidants can also prevent cellular damage, including to our energy-producing mitochondria, which are harmed by Parkinson's.[61] However, additional research is required to understand how diet might prevent the disease.

HELP YOURSELF TO ANOTHER CUP

Get brewing! Caffeine may be able to delay or prevent the onset of Parkinson's. Research has repeatedly shown that caffeine consumption is associated with a decreased risk of the disease.[62] And the more the better: a study in the *Journal of the American Medical Association* found that the incidence of the disease declined with increased amounts of coffee intake.[63] The effect is likely tied to caffeine rather than to coffee beans. Caffeine from other sources appears to offer a protective effect too, while decaffeinated coffee offers none.[64]

Ascherio and his colleagues, who studied the Mediterranean diet in those 130,000 health professionals, also looked at their caffeine consumption. After ten years, men who drank the most caffeine were 58% less likely to develop Parkinson's than those who consumed the least. Among women, moderate caffeine drinkers (one to three cups of coffee per day) had the lowest risk of developing the disease.[65]

Caffeine's actions in the brain may protect dopamine-producing nerve cells from the damage Parkinson's does to them.[66] When animals with Parkinson's symptoms are given drugs similar to caffeine, their motor function improves. In human studies, however, researchers have not replicated those benefits. This could be because by the time people are diagnosed with Parkinson's, too many nerve cells may already be lost—the caffeine prescription may come too late.[67] As suggested by Ascherio and others, prudent caffeine intake (one to four cups) early in life could protect nerve cells and reduce the risk of ever developing Parkinson's.[68] However, caffeine has its own costs. It can cause anxiety, headaches, and other side effects.[69] These effects must be weighed against its potential benefits for Parkinson's.

Taken together, all of us can lower our risk of Parkinson's by protecting our heads, exercising, eating a Mediterranean diet, and consuming caffeine.

TAKING CARE

Helping Those Who Bear the Parkinson's Burden

> One of the first things that happens when you get Parkinson's is the phone stops ringing.
>
> —Alan Leffler, retired business owner and Parkinson's disease caregiver[1]

TODAY, IN MANY PARTS OF THE WORLD, THE MAJORITY OF PEO-ple with Parkinson's are never diagnosed. Even among those who are diagnosed, most do not receive appropriate care. The almost universal underappreciation of this disease, however, stops at the doors of anyone who lives with or cares for someone with the disease.

Hans Jakobs, whose name has been changed, is a fifty-two-year-old bookkeeper from the oldest city in the Netherlands, Nijmegen. When Professor Bas Bloem, his neurologist and an author of this book, told him that the symptoms that had plagued him for years were caused by Parkinson's disease, he began to cry. So did his wife, Anna, and their daughter, Christine (names also changed). Bloem assumed that the family's tears were triggered by the bad news. Instead, the Jakobses were crying with relief.

A glass of water and several tissues later, Hans Jakobs was able to speak. He said, "Recognition, finally. The beast got a name."

Hans's dramatic story is, unfortunately, quite common.[2] Over the course of seven years, he had developed a series of symptoms that no one could explain. First, Hans, who had always been cheerful, became depressed. A family doctor prescribed antidepressants. They provided some relief, but Hans did not feel like his usual self. The doctor suggested it was some kind of burnout or a midlife crisis. But neither Hans nor his wife felt comfortable with this explanation. Their marriage was fine, and Hans really enjoyed his work where he had just been promoted.

Not long after Hans first started feeling depressed, Anna decided to sleep in another bed. That did not help his mood, but it was necessary. During his dreams, Hans would become violent and accidentally hurt her. "When he was fighting with animals that attacked him in his dreams, he would strike out," she said. "I consequently sustained bruises on a number of occasions so I had no other option."

As the years passed, Hans, who was in excellent physical shape, began to experience inexplicable fatigue. Modest bouts of exercise would send him to the couch. He visited an internist who ran some blood tests that came back normal.

"But I definitely knew that something was wrong when I also developed this sore shoulder on my left side," Hans said. "It was a complete mystery to me why this was happening since I am right-handed and barely use my left arm for heavy duty lifting. Yet that was the side that was aching all the time." Much to the family's frustration, the orthopedic surgeon whom Hans saw said that there was nothing wrong with the left shoulder. "Mark my words," the surgeon said, "this will be gone in a few weeks." But, of course, it was not.

His family doctor referred Hans to a physical therapist. The therapist, an experienced practitioner who was close to retirement, was the first to offer some hope in terms of finding an answer. "Your shoulder appears fine to me, but there is something funny about the way you walk. I am not sure what it is, but maybe you should see a neurologist instead of me."

To the experienced neurologist, the picture was clear. The doctor immediately noted his new patient's slowness in rising from his chair. Hans's face appeared friendly, but he did not blink at a normal clip. As he walked toward the office, his left arm barely moved. These were all unmistakable signs of Parkinson's.

In hindsight, the depressive spell, the acting out of his dreams, the fatigue, and the asymmetric shoulder pain were all early symptoms. There were other indications too, including loss of smell and constipation.

The false assurance Hans received that he was fine did not alleviate his suffering or his family's constant worry that something was wrong. Hans said, "Now that I know the name of the beast, I can finally begin to fight it."

SEEING WHAT IS IN FRONT OF US

Some conditions, like colon cancer, can remain hidden for years. Parkinson's, though, has so many visible features that when they are present, the diagnosis can readily be made.[3] Despite the multitude of clues, many are not diagnosed because patients or physicians erroneously attribute their symptoms to aging or fail to appreciate the issues.[4] In addition, some may not know that many aspects of Parkinson's are very treatable.

The number of cases that go undiagnosed—in both low- and high-income countries—is astonishing (**Figure 1**). In a 2003 study none of the individuals researchers identified in rural Bolivia as having Parkinson's had ever been diagnosed.[5] In the Italian Alps, 83% of those with the disease had not been told they had it.[6] In rural parts of China, 69% to 78% had not. In Beijing, almost half the people with Parkinson's were also undiagnosed.[7] And in southwestern Mississippi, the numbers were similar—42%.[8]

People with undiagnosed Parkinson's are all around us. In 2018, Samuel Jones (not his real name), a seventy-year-old retired mechanic, was hospitalized for possible depression in a small town in Alabama. Over the previous ten years, his walking had become slower, and his dexterity had diminished. He could no longer repair faucets or even hammer a nail. Jones, who had always been a jovial guy, also began interacting less with his friends and family. He became progressively withdrawn and spent more time in his favorite chair just watching television.

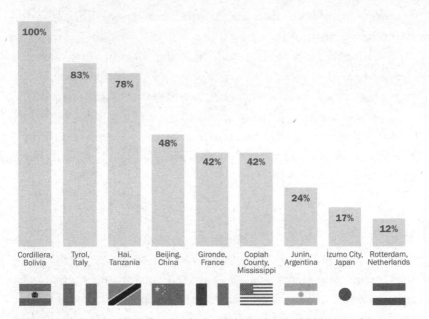

FIGURE 1. Percentage of individuals with Parkinson's disease who are undiagnosed by region.

His family grew concerned about his mood. They eventually took him in to the hospital. Fortunately, an astute psychiatrist noticed that Jones's main problem was not depression but slowed movements. Even though he did not have a tremor, it was clear that Jones, who could barely sit on a bed without support, had advanced Parkinson's disease. The doctors prescribed levodopa.

When the psychiatrist saw Jones again one month later, he asked how Jones was doing. Jones popped out of his chair, glided across the room, and smiled.

After a decade of needless suffering, Jones was fortunate to be diagnosed and treated. Many are not.

GETTING THE RIGHT TREATMENT

Once people are diagnosed, most do not receive the care they need. Treatment for Parkinson's rests on four pillars. Patients require appropriate medication, surgery in some instances, a multidisciplinary team of health-care providers, and, perhaps most importantly, their own personal support. The problem is that almost no one with Parkinson's gets all four.

In his fifties, John Harlan was a surveyor and an avid skier. Going down the slopes in the Adirondack Mountains in upstate New York, he noticed that something was off. His ease and grace seemed to be slipping away. He initially attributed the loss to age. However, he later became anxious and then developed cramping in his right hand. A local neurologist diagnosed him with Parkinson's disease and prescribed levodopa.

Like most, Harlan responded well to the medication and resumed an active life with his wife. But the drug's benefits did not last. The levodopa caused him to have involuntary movements that made him sway when sitting or standing. He could no longer be still.

A decade into the disease, his symptoms became more disabling. His feet would become stuck to the floor, or frozen, when he walked. Then Harlan's wife of forty years, who had helped care for him, died. Alone and struggling to shop, cook, and dress, Harlan decided to enter a nursing home.

There, Harlan received assistance but was unable to see his longtime neurologist. Instead, he was treated by an internist who had limited training in and experience with Parkinson's. Harlan's disease continued to progress. In an attempt to reduce his tremors and improve his other symptoms, the nursing home doctor added a new medication. Although well intended, the change made Harlan worse.

Harlan, who had held on to his fun-loving, congenial nature when he first moved to the nursing home, became confused and agitated. He physically struck out at nurses and fellow residents. The doctor started a medication to improve his memory and another to sedate him. Harlan only became more disoriented.

At one point, he climbed out a window, scaled a fence, and tried to leave the nursing home. After he was found and led back to the facility, the staff placed him in restraints. Harlan's confusion and despair worsened.

An innovative administrator at the nursing home reached out to Parkinson's specialists at a major medical center three hours away in Rochester, New York. He asked if they would be willing to see Harlan via "telemedicine"—essentially doctor visits over video conferencing. At the time, in 2007, video conferencing technology was in its infancy, telemedicine was novel, and using it in nursing homes was almost unheard of. Fortunately, an open-minded specialist said yes.

In his initial video visit, Harlan had slurred speech, writhing movements, and frequent episodes of freezing. He also exhibited signs of dementia. He scored twenty-one out of a possible thirty on a basic cognitive test. Over the next nine months, Harlan "saw" his Parkinson's specialist six times without ever having to travel. At each visit, Harlan's nurse accompanied him. She supplemented Harlan's story with her own observations.

Based on what he heard and saw, the specialist eliminated two medications and adjusted the timing and dosage of others. Harlan's confusion and dementia disappeared; his score on the cognitive test improved to a perfect thirty. The doctor also referred Harlan to physical therapy at the nursing home. Harlan's mobility and confidence improved. Although his days on the slopes were long behind him, Harlan was ready to take walks again.

Thanks to this team approach, Harlan gradually regained his personality and his freedom. He never hit anyone again and lived three more pleasant years. At age eighty, Harlan died of advanced Parkinson's disease.

Harlan's story is exceptional in that he was eventually seen and treated by a specialist. Few are. Two-thirds of people with Parkinson's in US nursing homes never see a neurologist in the clinic, and only a handful of these specialists go to nursing homes.[9] Ninety percent of residents do not receive physical therapy. Fewer than 10% see a mental health specialist.[10]

Other countries have similar shortcomings. In the Netherlands, many nursing home residents with Parkinson's do not receive the right medicines. Some receive inappropriate doses. Others are prescribed drugs that worsen the disease.[11] Almost 85% die within three years, which may be due to the advanced stage of the disease. But many nursing home residents do not receive the adequate care and support they deserve in the final years of their lives.[12]

Up to 40% of people with Parkinson's will end up in nursing homes and frequently receive sub-standard care. In addition, many people develop Parkinson's disease after moving to a nursing home and are never diagnosed. They experience considerable and at least partially avoidable disability.[13]

One of these patients was a seventy-eight-year-old man who had lived in a nursing home for several years and become bound to his chair and bed. A visiting neurologist saw obvious signs of advanced Parkinson's. He recommended levodopa. The improvement was incredible. The man started walking again, and his mobility increased such that he could leave the nursing home to shop without assistance. As part of a standard cognitive test, the patient was asked to write a simple sentence (**Figure 2**). He wrote, "The sun has started to shine again in my life."

Even the newly diagnosed have limited access to care. Dr. Allison Willis, now at the University of Pennsylvania, and her colleagues examined the records of 130,000 Medicare beneficiaries who had recently

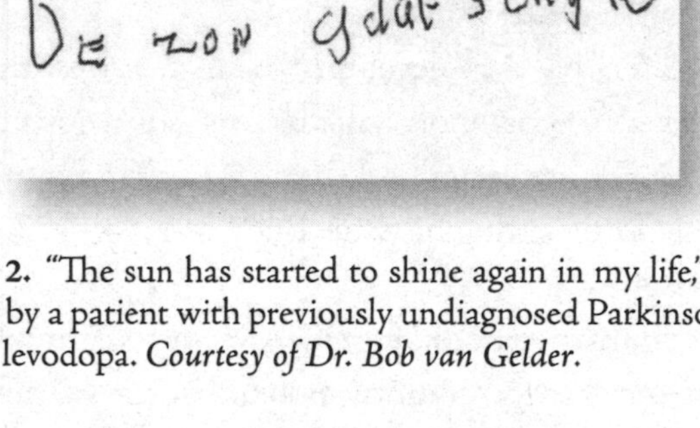

FIGURE 2. "The sun has started to shine again in my life," written in Dutch by a patient with previously undiagnosed Parkinson's after receiving levodopa. *Courtesy of Dr. Bob van Gelder.*

been diagnosed with Parkinson's disease.[14] They found that only 58% had seen a neurologist within four years.[15] Those who had not were 16% more likely to fracture their hip, 27% more likely to be placed in a nursing facility, and 28% more likely to die prematurely.[16]

The United States spends 40% more on health care per person than any other country.[17] What are we paying for if almost half of our seniors with a neurological disorder do not see a neurologist? African Americans, women, and the elderly with Parkinson's disease are even less likely to see a neurologist.[18] As just one example of the treatment disparities, among 8,400 individuals who have received deep brain stimulation for Parkinson's, only 88 (1%) were African American.[19]

Appropriate neurological care improves health and saves money. In the United States, individuals with Parkinson's who see neurologists have fewer hospitalizations and lower health-care costs.[20] The more frequent the visits, the lower the costs.

RECOGNIZING THE TOLL

Not only is Parkinson's the fastest-growing neurological condition on the planet, but it is also among the most disabling.[21] Individuals with Parkinson's disease suffer. And so do their caregivers.

Dr. Bob Dein is a successful, mild-mannered pathologist from Florida. In 2002, his wife, Barbara, was diagnosed with Parkinson's. At that time, they had been married for thirty-two years. Barbara, ashamed and fearful of her disease, asked her husband not to tell anyone of the diagnosis, including their grown son.

A decade later, Barbara developed hallucinations and started to become confused and paranoid. She became suspicious that her husband would betray her secret and that her husband was seeing other women. Afraid of upsetting his wife, Bob hid his wife's new psychosis even from her physicians.[22]

The disease burdened both of them. After years of suffering without any support from their community, Bob did not see a way out. During one especially bad night, his wife began throwing things and slapped her medication out of his hand. Bob, out of exasperation, threw a glass of water in her face.

He was horrified by what he had done. The next day, he decided to end his life. He had already come up with a detailed suicide plan. He would drive his boat far out into the Gulf of Mexico. He would then tie two car batteries to his ankles and jump into the water.

His wife had an appointment that morning. After dropping her off, Bob waved good-bye, went home, and started up his boat. But he could not bring himself to leave. Barbara had looked so weak. Eventually, he went to pick her up from her appointment, but he had made an important decision—he would seek help.

He called her neurologist, who agreed to admit Barbara to a local hospital to adjust her medications under close observation. While hospitalized, she had a psychotic episode that required an entire hospital wing to be locked down. Under the state legal system, Barbara had to be transported to a psychiatric hospital for mandatory care. She never came home, and two years later, she died in a memory-care unit. The legal system had done what Bob could not do himself—and possibly saved his life.

More than 34 million Americans care for an adult over fifty; almost half of these adults have a neurodegenerative disorder like Parkinson's disease.[23] These caregivers suffer in various ways. In a British survey of caregivers, over 40% (most of them women) reported their health had worsened because of this work. Almost half had elevated depression scores, and two-thirds felt that their social lives had suffered.[24] In the United States, older people who take care of another adult are 60% more likely to die sooner than noncaregivers. Many, like Dr. Bob Dein, contemplate suicide.[25]

After his wife died, Bob began taking photographs of people with the disease. **Figure 3** is an early photograph of his wife. Bob and Barbara's story illustrates the tremendous burden that Parkinson's can be. It is also a powerful tribute to the wonderful resilience shown by

FIGURE 3. Photograph of the late Barbara Dein. *Courtesy of Dr. Bob Dein.*

families who have to cope with Parkinson's. Like Bob, many spouses and other caregivers show amazing commitment to caring for their loved ones. And Bob is not alone in turning this ordeal into something meaningful. In his case, he is using photography to depict the many different faces of Parkinson's to educate and inspire others.

One reason why Parkinson's is such a devastating disease is that people can experience a vast array of disabling symptoms. As we've seen, the toll goes beyond the tremors most of us associate with it.

Stephanie Hughes, who also asked that her name be changed, shows how far the disease can go. She is a mother of three and a grandmother of two. She is also the wife of Jack Hughes (also not his name), a seventy-one-year-old Vietnam veteran and former grocery store manager. Married for nearly forty years, the couple has lived what Stephanie calls a "blessed" life in upstate New York. But six years ago, Jack developed a tremor in his lip and would often find himself drooling.

He had been exposed to Agent Orange in Vietnam. "It was everywhere," he said. After his diagnosis, he started on levodopa, and his symptoms were initially well controlled. The couple continued to be active, traveling throughout the country and around the world.

But as the disease progressed, so did the severity of Jack's symptoms. He also developed new ones, including an overactive bladder, which is common in Parkinson's disease. He initially had a few accidents but is now incontinent. At night, he has to wear adult diapers, but they do not work well.

Each night, Stephanie lays out two bed pads, a highly absorbent towel, and an extra sheet over the area where her husband sleeps. Most of the time that is sufficient, but on some nights Jack soaks through all of them. Not surprisingly, both wake up exhausted and frustrated. Jack regularly falls asleep at unwanted moments, including in the middle of conversations with his friends or Stephanie.

Parkinson's has also affected Jack's ability to swallow. He began to lose weight and to have trouble taking his medications. So a few years ago, doctors placed a feeding tube into his stomach.

Most mornings, Stephanie makes Jack pancakes, which is one of the few things he can still eat. As the day goes on, it becomes harder for him to swallow. Stephanie then gives him extra calories through his stomach "plug." Despite her best efforts, he continues to lose weight.

Over the last year, Jack's thinking has worsened. Most days, he can still carry on conversations with his wife. Recently, though, he has started speaking less, and Stephanie does most of the talking for him. When he visits the doctor, he is quiet. He cannot name the month or the president.

Because his Parkinson's stems from his exposure to Agent Orange, the Hughes are entitled to two hours of home health care daily through the Veterans Health Administration. But there is a shortage of home health aides in their area, so they are not always able to get the help they need and earned.

Stephanie is healthy and still faring well, but on some days she is overwhelmed. She will call her children to see if they can lend a hand.

Asked if they are bitter about their circumstances, Stephanie replies, "No. After my husband first retired, we had a good five years to enjoy life and travel. Not everyone gets that."

IT TAKES A TEAM

As heroic as caregivers can be and as worthwhile as it is to see a neurologist, they are often not enough. Even levodopa and other medicines have their shortcomings. Not all features of Parkinson's respond to drugs.[26]

People with Parkinson's often benefit from a team of specialists. The disease has so many symptoms that professionals from thirty different disciplines, including dentists, dieticians, pharmacists, and psychiatrists, can provide help. Not every patient will require every specialist, and many experts will only be needed temporarily. Ideally, the team is tailored to an individual's needs.

Frans de Wit (not his real name) is a Dutch plumber with Parkinson's. His experience exemplifies the limitations of the generic medical care that many patients receive. When diagnosed in his fifties, he was started on levodopa and was pleased by his improvement. He even resumed working.

After about seven years of treatment, though, Frans began to experience sudden and unexpected episodes during which his feet felt as if they were glued to the floor, a common occurrence in Parkinson's. These episodes were initially relatively innocent. But he and his wife, Marijke (name changed), began to worry when these freezing episodes started leading to falls. After Frans broke his wrist, his neurologist decided to increase his levodopa dose. His symptoms improved but only temporarily.

Frans's walking problems became worse. He rarely went outside. Fearful of another accident, he restricted his walking even within his own home. His internist referred him to a physical therapist who had previously treated Frans for low back pain. The therapist had limited experience with Parkinson's but tried putting Frans on a treadmill twice a week. This regimen did little to alleviate Frans's freezing episodes, and his mobility decreased further.

Then the de Wits visited a "Parkinson Café." Organized by the Dutch Parkinson Association, these cafés are actually support groups whose name offers cover due to the stigma that often accompanies the disease. At the café, the de Wits learned from fellow patients and caregivers about ParkinsonNet, a network of health professionals with extensive experience and expertise in Parkinson's care (**Box A**).

To be a part of the network, clinicians, including physical therapists, nutritionists, social workers, and sexologists, go through intensive

---

BOX A. WHAT IS PARKINSONNET?

ParkinsonNet is a new model of health-care delivery that seeks to provide optimal care to patients with Parkinson's. The creation of Dr. Marten Munneke, a physical therapist and epidemiologist, and Professor Bas Bloem, a Dutch neurologist and one of our authors, it was developed in 2004 to train physical therapists in several regions in the Netherlands. Since that time, ParkinsonNet has grown to seventy regional networks and trained 3,400 professionals from nineteen different disciplines. The goal is to empower patients with knowledge about their disease and connect them to an array of highly trained Parkinson's specialists whose experience is enriched by seeing many people with the disease.

Better care through ParkinsonNet has led to better outcomes and lower costs. One study found that the care model reduced hip fractures by 50% and decreased hospital admissions. Numerous studies have demonstrated the network's benefits for patients.[27] Because of ParkinsonNet's success, the program has now expanded to California (through Kaiser Permanente), Luxembourg, and Norway. Other countries are likely to follow soon.[28]

in-person training and continue with regular follow-up courses. Patients can then find trained therapists near them through a website. Through the network, patients have also formed their own regional communities where they advise researchers on the best study designs for their population.[29] An educational television series called ParkinsonTV stars people with Parkinson's and offers more information. It is now available online at www.parkinsontv.org.

Based on recommendations from his new friends, Frans went to see Trudy Bloem (no relationship to Bas Bloem), a ParkinsonNet physical therapist. As soon as he entered her clinic, he noticed a difference in the experience and the expertise. He saw several other individuals with Parkinson's exercising. Some were walking on treadmills; others were cycling on stationary bicycles.

Bloem's approach was different from the start. She was already familiar with the freezing and many of Frans's other symptoms. She also had ready solutions.

She tested whether walking paced by the rhythmic sound of a metronome might help Frans. She knew that audio cues could help people with Parkinson's maintain rhythm when walking and avoid freezing. Much to Frans's surprise, it did. Finding the optimal frequency required a bit of tweaking. Although his freezing was not cured, his walking and balance were better. In another test, Frans stepped over stripes that were separated by small distances on the floor. By reminding him to lift his legs higher, these visual cues also helped his walking.

Frans now uses small music ear buds, which his grandson gave him, to listen to the beat of the metronome and improve his walking, especially outside. At home, he has worked together with Bloem to paste white stripes onto the floor in critical places, such as near doorways. As a result, he feels more independent and falls much less.

REACHING THE COMMUNITY

ParkinsonNet has successfully trained professionals in many, predominately urban areas. However, most of the world lacks access to such care. In rural areas and in some lower-income countries, though, there are simply not enough experts. In India, nearly 1 billion people live in areas without a practicing neurologist. For twenty-three sub-Saharan African countries, the average population per neurologist exceeds 5 million. For example, only three neurologists serve a population of around 50 million in Tanzania. In the Arab world, there is only one neurologist for every 300,000 individuals.[30] Training new physicians and neurologists is necessary but will require considerable resources and time.

Technology may help bridge the gap. One model for how this can work is the Extension for Community Healthcare Outcomes (Project ECHO).[31] Project ECHO was developed in 2011 to support primary care physicians caring for individuals with hepatitis C in rural New Mexico. It connects local clinicians to remote experts via video conferencing. During weekly or monthly sessions, clinicians learn more about the disease and discuss with experts how best to care for their local patients who require more specialized care.

Project ECHO now covers fifty conditions, from diabetes to dementia, and reaches twenty-one countries on six continents.[32] Extending Project ECHO to Parkinson's could improve care in areas of the world with few neurologists and increase the capacity of clinicians who are already seeing the majority of individuals with the disease.[33] The initiative could also be extended to engage other key clinicians, including nurses and therapists.

TAKING CARE TO THE PATIENTS

Both ParkinsonNet and Project ECHO bring expert care into communities that previously lacked it. Other programs do the same by returning care to the home.

Current care for Parkinson's could not be designed worse. At its best, it requires caregivers to drive often elderly individuals into major urban centers for doctor's appointments. These visits provide physicians with superficial snapshots of complex and fluctuating symptoms and leave nearly all participants overburdened and overwhelmed. Placing the burden on patients means that those with the greatest need often have the least access to care.

In the 1930s, the house call brought care to patients and accounted for 40% of all doctor visits.[34] Since then, house calls have almost disappeared. At their nadir, less than 1% of Medicare patients received such care.[35] Fortunately, they are beginning to make a comeback.[36]

Dr. Jori Fleisher, a Parkinson's specialist now at Rush University in Chicago, and her colleagues are bringing back the house call. A specialist, social worker, and nurse visit individuals with Parkinson's in their own homes. Over two years, they have conducted nearly three hundred home visits to eighty-five individuals. The program, funded by the Edmond J. Safra Philanthropic Foundation and the Parkinson's Foundation, has reached people with the most advanced disease whose average age is eighty.[37] The seventy-minute visits include conversations with the social worker and caregiver, a detailed medication review by a nurse, and the development of a comprehensive care program that is discussed with the patient and the caregiver. Most visits result in a change to medication or additional treatment recommendations, and satisfaction with the program is high among patients and caregivers.[38]

The University of Florida launched a similar initiative called Operation House Call in 2011. The program, funded by the Smallwood Foundation, sends Parkinson's disease fellows—specialists in their last year of medical training—to visit patients in their homes. Many live in rural areas, lack health insurance—some are too young for Medicare—or cannot afford to travel. The fellows listen to patients, perform examinations, and provide recommendations to improve their health. As a result, patients' symptoms have improved, and fewer hospitalizations have occurred.[39]

The same video conferencing that helped John Harlan in the nursing home is also enabling virtual house calls. Not surprisingly, patients like receiving care closer to home.[40] These visits eliminate the waiting room, the parking garage, and driving, which is impaired in Parkinson's disease.[41] On average, this next generation of house calls saves patients three hours of time and one hundred miles of travel per visit.[42] Some patients even say that virtual visits offer a more personal connection than traditional appointments.[43]

The roots of telemedicine for Parkinson's began in Kansas in 1992.[44] A pioneering neurologist wanted to reach rural Kansans who had the disease and address the inequality in medical care. Most people in Kansas do not live near a specialist. So Dr. Jean Hubble, then at the University of Kansas, came up with a novel solution. She used a new technology called video conferencing to connect to satellite clinics throughout the state. She then could see patients who lived hundreds of miles away without their having to travel long distances.

Hubble was uncertain of how patients would respond. She wondered if the patients would be comfortable in front of a video camera, if they would feel that they had truly seen a physician, and if they would miss their doctor's touch. But the patients liked telemedicine, and all thought that the videoconferencing improved access to good care.[45] Since then, telemedicine for Parkinson's disease has begun to gain broader acceptance and use.[46]

Future care could marry the ability to see experts remotely with smartphones and wrist sensors that can monitor symptoms. These technologies allow for passive monitoring in a person's natural environment. Recent work has shown that, with appropriate support, older adults are willing and able to use such sensors.[47] In addition, these approaches can detect subtle changes that are invisible even to a trained eye.[48]

In the future, technology could enable us to reach those huge populations of people who are undiagnosed and, as a result, not receiving treatment. A smartphone app may even facilitate diagnosis of the disease. Individuals might be able to tap an app or use a smartwatch to assess their movements to determine if they might have Parkinson's. They could then use the same device to reach a clinician who could conduct a remote exam and provide care.[49]

Such applications will be especially important in middle- and low-income nations. In countries like China and India, physicians are rare, but smartphones are plentiful. Some of these applications are beginning to emerge, including for the evaluation of stroke in rural India.[50] The time when a phone, watch, or other digital device can inform users of their Parkinson's disease status and eventually connect them to care may be closer than we think.[51]

REMOVING ECONOMIC BARRIERS TO CARE

Given these effective models, why are so many people with Parkinson's still going without good care? In the United States, the answer is, in part, Medicare's policies.

Created in 1965, the universal health insurance program for people sixty-five and older is one of the great policy accomplishments of the last century. At the time it was created, only half of older adults had access to medical care.[52] Now, almost all do—and life expectancy for those over sixty-five has increased.[53]

But in the five-plus decades since its inception, Medicare has not kept up with our shifting medical problems. When the program started, the big health issues were acute conditions, like heart attacks and strokes. But the principal challenges for the current generation of Medicare beneficiaries are not these emergencies but chronic illnesses like Parkinson's. Because rates of Parkinson's increase with age, most Americans with the disease are insured through Medicare. This is where the program falls short.[54]

Most older adults want to live in their homes or communities for as long as possible.[55] And yet, 88% of Parkinson's disease health-care costs are spent on hospitals and nursing facilities.[56] That is because Medicare's financial incentives direct care away from peoples' homes and communities and toward expensive institutions.[57] For example, Medicare's reimbursement for care provided in the community is modest. It pays about $100 for a thirty-minute follow-up appointment for Parkinson's disease. If the same doctor visit occurs in a hospital-based clinic, Medicare pays double—$200.[58]

Worse, Medicare pays even more for complications of the disease, many of which are preventable. For example, if a man with Parkinson's falls and breaks his hip, Medicare will pay approximately $25,000 for surgery and institutional care.[59] Of that $25,000, about 85% goes to the hospital or skilled nursing facility following hospital discharge. In studies that will surprise few, hospital care for individuals with Parkinson's is frequently unsafe.[60] Patients encounter delayed treatment, the wrong medications, prolonged immobility, lengthy stays, and high mortality.[61]

The $25,000 for the broken hip is one hundred to two hundred times what Medicare pays for a visit to a neurologist to optimize balance and overall function, which we know can help prevent falls.[62] In short, Medicare pays little for preventive care and much for complications. And the situation is not much different in other countries. In England, the National Health Service pays around £29 million on emergency admissions following falls by individuals with Parkinson's and almost £3 million for preventable bladder infections.[63]

Not surprisingly, individuals with Parkinson's fall a lot. More than 60% have fallen at least once. Almost 40% fall repeatedly. In the latter group, the average number of falls per year is twenty.[64] Many of these falls result in hip fractures, which are two to four times more common among those with Parkinson's compared to similarly aged individuals without the disease.[65] Lessons for learning tai chi, which can prevent falls, are not covered by Medicare.[66]

In addition to the resulting pain and disability, the economic costs of falls are staggering. One US study estimated that the medical costs—principally to health insurers—was $50 billion annually.[67] Of that, $38 billion—more than the entire research budget of the National Institutes of Health—is paid by taxpayers through Medicaid and Medicare.[68]

Long-term care facilities, such as nursing homes, now house one-quarter of Medicare beneficiaries with Parkinson's.[69] Medicaid, which is the largest payer for nursing home care, spends about $80,000 per person per year for such care.[70] These resources could be better directed toward less expensive home care; yet Medicare only provides such coverage in limited circumstances, such as after discharge from a hospital. In essence, Medicare and other public funders provide the most coverage—and pay the most—for services that patients seek to avoid and the least for services that patients and their caregivers want and need. As a result, affected individuals and their families suffer.

Telemedicine could also save Medicare money and provide patients better care—if only Medicare would cover it. The Veterans Health Administration has embraced virtual visits.[71] Consequently, veterans throughout the country receive care remotely from Parkinson's specialists and even mental health professionals.[72] However, Medicare's coverage of telemedicine is limited to clinical settings (e.g., hospitals) in "health professional shortage areas," which are typically rural counties. Care provided in the home via telemedicine is not covered at all.[73] In 2016, Medicare spent little—less than $30 million, or less than 0.01% of its nearly $600 billion budget—on telemedicine.[74]

Unlike so many other countries, the United States is fortunate to have enough general neurologists to care for individuals with Parkinson's disease (**Figure 4**).[75] You would think it would be possible for more patients to reach them.

Medicare needs reform to improve care and lower costs. Currently, a committee created by the American Medical Association helps set Medicare reimbursement rates.[76] The process is anything but open and democratic. The meetings are generally closed, most members have no term limits, the proceedings are private, and minutes were only recently made public.[77]

The committee consists exclusively of physicians, and representation is linked not to their care of Medicare beneficiaries but rather to their specialties.[78] The result is that smaller groups, such as plastic surgeons, have an equal voice to larger groups, such as family physicians.[79] The smaller, but more numerous, groups tend to do things to patients, while the larger groups tend to care for patients. The committee thus favors higher rates for those who perform procedures.

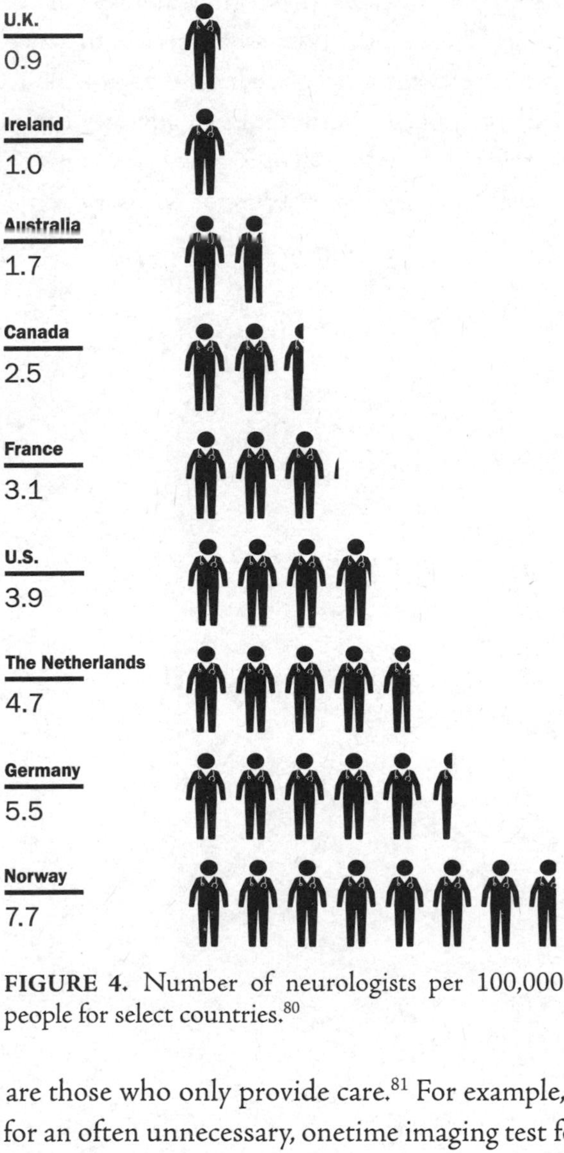

FIGURE 4. Number of neurologists per 100,000 people for select countries.[80]

The losers are those who only provide care.[81] For example, Medicare pays more for an often unnecessary, onetime imaging test for Parkinson's disease than for caring for someone with the disease for four years.[82]

The committee, like those that advise the Food and Drug Administration (FDA) on approval of drugs and devices, should include patients.[83] Some patients have contributed taxes to Medicare for the last fifty years. In many cases, these individuals would be far better positioned to make decisions about what services they need and what offers true value to them than physicians with inherent conflicts of interest.[84] Patients gained a voice on FDA advisory committees after a report commissioned by President Richard Nixon made the recommendation.[85] The time is long overdue for Medicare to take the same step forward.

8

HOPE ON THE HORIZON

The Promise of New Treatments

The latest research is the latest hope.

—Michael J. Fox

THERAPEUTIC PROGRESS AGAINST PARKINSON'S HAS BEEN SLOW. In the last twenty years, the Food and Drug Administration (FDA) has approved about a dozen new drugs for Parkinson's (**Table 1**).[1] Nearly all of them, however, are only incremental advances on existing therapies. Three of the new drugs are just different formulations of levodopa, the dopamine precursor.[2] Others treat the increased involuntary movements that levodopa causes. Some decrease the symptom fluctuations that can develop after long-term use of the drug—there can be periods when levodopa works well and other times when it does not.

In this century, only two new drug classes—therapies that take a different approach to tackling the disease—have come on the US market. However, both only treat symptoms that affect a subset of individuals: pimavanserin for hallucinations and droxidopa for light-headedness. Fortunately, in addition to the gene-targeting treatments likely on the way, innovative surgeries, possible immunizations, and other new therapies are emerging.

YEAR	DRUG	INDICATION	NEW CLASS OF MEDICATION
2018	Levodopa inhalation (Inbrija)	Fluctuations of symptoms that can develop after use of levodopa[3]	
2017	Amantadine (Gocovri)	Involuntary movements due to levodopa[4]	
2017	Safinamide (Xadago)	Supplemental therapy to levodopa to treat fluctuations[5]	
2016	Pimavanserin (Nuplazid)	Hallucinations and delusions associated with Parkinson's disease[6]	✓
2015	Carbidopa/levodopa intestinal form (Duopa)	Fluctuations of symptoms in advanced Parkinson's disease[7]	
2015	Carbidopa/levodopa extended release (Rytary)	Treatment of Parkinson's disease[8]	
2014	Droxidopa (Northera)	Lightheadedness when standing[9]	✓
2007	Rivastigmine (Exelon)	Dementia associated with Parkinson's disease[10]	
2007	Rotigotine (Neupro)	Treatment of Parkinson's disease[11]	
2004	Apomorphine (Apokyn)	Fluctuations of symptoms associated with advanced Parkinson's disease[12]	
1999	Entacapone (Comtan)	Supplemental therapy to levodopa to treat fluctuations[13]	

TABLE 1. New FDA-approved drugs for Parkinson's disease, 1999–2018.[14] Note: Other formulations of existing drugs have also been approved during this period.

THE POTENTIAL OF DEEP BRAIN STIMULATION

Frida Falcon, a successful Peruvian sculptor in her fifties, and her husband, Marcus, a multinational businessman, were enjoying a good life together. Then "Mr. Parkinson," as they came to refer to the disease, knocked at their door. As the illness took increasing control over Frida, her ease of movement and artistry disappeared. Medication helped but only temporarily. When her levodopa kicked in, her tremors would fade away. But over time she needed to take the drug every hour.

Then she developed new, more disabling movements. These writhing motions prevented her from sculpting or even sitting still. The more levodopa she took, the worse the movements became. When she took less, she stiffened and moved more slowly.

The Falcons were fortunate enough to have the resources to fly around the world and meet with different doctors. Eventually, they decided to pursue deep brain stimulation surgery at the University of Florida under the guidance and care of one of our authors, Dr. Michael Okun, and his colleague, Dr. Kelly Foote. During a four-hour procedure that requires the patient to be awake, wires that produce electrical impulses are implanted in the brain. The thought of sticking metal into her head was obviously not appealing to Frida, but in select individuals, the surgery can reduce the involuntary movements caused by levodopa, decrease the symptoms of Parkinson's, and improve quality of life.[15]

Surgery as a treatment option for Parkinson's first emerged over one hundred years ago.[16] Doctors noticed on rare occasions that a stroke or brain tumor in specific locations—parts connected to the substantia nigra where dopamine-producing nerve cells are located—could relieve the symptoms of Parkinson's. Based on these observations, surgeons in the 1930s and 1940s began damaging (usually by burning with an electric current) specific areas of the brain to try to alleviate tremors. Not surprisingly, slight errors had profound consequences. Missing the target by even a millimeter or two led to visual loss, problems with speech, or paralysis. When the surgeons hit their target, a patient's tremor, though, could magically disappear. With the arrival in the 1960s of levodopa, which was safe and effective, risky surgeries for Parkinson's soon fell out of favor.[17]

In the 1970s, interest in surgery was renewed when Dr. Mahlon DeLong, a gifted researcher at Johns Hopkins, began to map the brain circuits important to Parkinson's disease. DeLong created a model of the brain circuitry that showed connections between the substantia nigra and many other parts of the brain.

DeLong identified brain areas that were responsible for movement. Other brain circuits are responsible for seeing or speaking, for example. With a model of the movement network, DeLong and other scientists could then predict what might happen from a loss of nerve cells in the substantia nigra. This loss would decrease its output and affect other parts of the brain circuit that then might become under- or overactive.[18] These areas could then be targets for surgical treatment that would usually decrease the output from regions that were hyperactive.

With DeLong's map of the brain circuits, surgeons could now explain why earlier surgeries that quieted certain targets worked. They could also surmise why other surgeries did not. The time was ready for a more refined surgical approach. In 1987, Dr. Alim Benabid in Grenoble, France, came up with one. Rather than destroying nerve centers, he suggested using electrical current to change their output. This approach was called deep brain stimulation.[19]

This new surgery uses a small electricity-conducting wire to stimulate a targeted area of the brain. The wire is connected to a battery that is placed under the skin on a patient's chest much like a pacemaker for the heart. The battery-powered wire can then provide electrical current to change the nerve cell activity in the targeted region. This stimulation actually decreases or blocks the output of the target.[20]

However, rather than irreversibly destroying the nerve cells like earlier surgeries, this stimulation can be modified based on a patient's symptoms. Like a television, the stimulator can be adjusted with a wireless remote control device. In the clinic, when a patient's stimulator is first turned on, the results can be dramatic. Tremors can vanish, stiffness can disappear, walking can become normal, and hidden smiles can reappear.

Over time, Parkinson's progresses, and adjustments to the deep brain stimulator are required. And unfortunately, over enough time the miraculous effects tend to disappear. Still, Dr. David Marsden, the late British neurologist and one of the founders of the movement disorders field, called deep brain stimulation one of the two miracles in Parkinson's disease (levodopa was the other).

Numerous studies have demonstrated the benefits of the surgery, which has become part of standard Parkinson's care in many regions of the world.[21] There are risks, including bleeding, infection, misplacement of the electrode, and, very rarely, death. To date, over 200,000 patients worldwide, including Frida Falcon, have received the surgery.

For Frida, the surgery was a success. She is back to sculpting and jokes that she has divorced "Mr. Parkinson." The improvement in her movements and quality of life are not permanent but can last from five to ten years.[22] She is among the lucky ones.

Many people in most of the world lack access to these expensive surgical techniques—even in the United States. In 1997, the FDA approved the surgery for the treatment of Parkinson's disease that is not well controlled by medications.[23] Deep brain stimulation costs about $65,000, which is generally covered by insurance and Medicare. But the surgery is primarily performed at major medical centers—in the United States and in other high-income nations—which often leaves people living in rural areas at a disadvantage.[24] Those who do have access to the surgery usually have to go back every few months to have their stimulators adjusted. Doctors reprogram the stimulators— almost like changing the television channel—to find the settings that best relieve a patient's symptoms.

New advances are enabling physicians to make these adjustments remotely, which, as it sounds, does not require the patient to travel to the medical center. Doctors in China are already doing this while video conferencing with their patients. The physician cycles through many possible settings until the neurologist and patient find an optimal one that, for example, helps minimize tremors while avoiding side effects from excessive stimulation.

In the future, patients will be able to wear a watch or sensors that can passively monitor their movements. Data from these sensors will be automatically sent to clinicians who can use the information to adjust medications or stimulator settings. The information may even be able to bypass the physician altogether. Like glucose sensors for insulin pumps, the data could guide the stimulator to optimal settings directly. Studies testing this technology have already begun, and Okun along with Dr. Christopher Butson at the University of Utah have been working with the National Institutes of Health to make home-based adjustments in Parkinson's devices a reality.[25]

Also on the horizon are more sophisticated stimulators that will be able to predict, rather than react to, tremors. Dr. Helen Brontë-Stewart, a neurologist at Stanford University, and Dr. Peter Brown, a neurologist at the University of Oxford, are now using the wire electrode not only to send electrical impulses but also to sense abnormalities in a brain's natural rhythm. These blips can indicate the onset of a tremor or can identify when someone's walking is about to be interrupted by a freezing episode. In two proof-of-concept studies, Brontë-Stewart and her colleagues demonstrated that this adaptive form of deep brain stimulation that responds to a patient's own signaling can reduce tremors and improve walking.[26] So symptoms can be prevented before they even occur.

TURNING CELLS ON WITH A FLICK OF A SWITCH

The emerging field of optogenetics offers more promise for Parkinson's. As the name suggests, the idea is to activate certain genes and the proteins they command using light ("opto"). The work is still in very early stages and has only been tested on animals, but the approach is fascinating.[27]

Here is how it works. A virus is used to insert specific genes into nerve cells in the brain. The most important of these genes tells the body to make a protein that turns on—starts working—in response to light. In the 1970s, researchers found this unique light-activated protein in the tiniest of organisms.[28] This class of small living creatures is found in very salty environments, such as the Dead Sea. Most remarkably, unlike almost all organisms, these creatures do not require sugar for energy. Instead, they can use light.[29]

Putting a light inside a human brain is not yet feasible. However, scientists can use light to study the circuitry that DeLong modeled in the brains of animals with Parkinson's.[30] A fiber-optic light source is either mounted onto the skull or placed deep within the brain. When the light is turned on, so is the activity of the protein in the cell. Turning on the light turns on the protein. Turning off the light turns off the protein.

The proteins that are turned on or off by the light control the activity or electrical firing of the nerve cell. Light can activate or excite these now light-sensitive cells. The scientists can see what happens in the circuits that are responsible for movement as well as speech or thinking.[31] They can test to see whether activating certain nerve centers in the brain improves or worsens the gait of a parkinsonian mouse all with just light. In essence, they can try to mimic the effects of deep brain stimulation without the surgery.

Researchers at Stanford used light to study the function of different nerve cells in animal models of Parkinson's disease. Nerve cells that had the light-sensitive protein were turned on or off at the discretion of the researchers. They found that light, depending on which cells it turned on, could worsen or improve Parkinson's symptoms in mice.[32] For example, shining the light on certain nerve cells "elicited a parkinsonian state"—the mice's movements and walking were slowed.[33] By contrast, using a light to turn on dopamine-producing nerve cells "completely rescued" animals from their parkinsonian symptoms.[34] In essence, light, and not a medication like levodopa or a surgery like deep brain stimulation, could restore function to animals with Parkinson's disease.

In the short term, light can help us understand how nerve cells control movement. In the long term, the hope is that surgical treatment with light or optogenetically inspired therapies could alleviate the symptoms of Parkinson's disease.

TARGETING GENES TO TREAT THE UNDERLYING CAUSE

In 2008, Patti Meese, an effervescent fifty-six-year-old human resources director, developed a shuffling gait and stooped posture and "was pretty much a mess," she said, when she was diagnosed with Parkinson's. Receiving that news sent her into what she later described as a "huge depression." She went into isolation, which she now says is the worst thing that you can do with the disease.

Patti Meese, though, is not one to let anything keep her down. She rebounded from her depression and started doing water aerobics and using a stationary bicycle for up to two hours a day. Soon, she entered El Tour de Tucson, Arizona's largest bicycling event. To raise money and awareness, she formed her own group of competitors, called "Team Foxy Heels," and completed a ten-mile ride in six-inch stilettos (**Figure 1**). She also started volunteering for every research study that she could. By her count, she has participated in more than one hundred in the ten years since her diagnosis.

From one of the studies, she learned that, like Google cofounder Sergey Brin, she carried a mutation in the LRRK2 gene, which can

FIGURE 1. Patti Meese in six-inch heels at the 2011 El Tour Fun Ride in Tucson, Arizona. *Courtesy of Patti Meese.*

raise the risk of developing Parkinson's by 20% or more.[35] Mutations in the LRRK2 gene increase the activity of the LRRK2 protein.[36] Researchers are developing new therapies to reduce its activity.

In an early study of one such drug, researchers at Denali Therapeutics, a biotechnology company in Silicon Valley, found that the drug was safe and well tolerated in healthy volunteers. In the fluid surrounding the brain, high levels of the drug were found, suggesting that the drug is able to get where it needs to go.[37] The next step is to test the drug in people with Parkinson's to make sure it is safe—work that is now underway.[38]

Denali is not alone in pursuing treatments aimed at LRRK2. Multiple companies, including Biogen, Cerevel Therapeutics, GlaxoSmithKline, and Merck, have all created drugs that target LRRK2.[39] For the 1% to 2% of individuals who have Parkinson's due to mutations in the LRRK2 gene, the drug could slow the progress of the disease.[40] For those who carry a mutation but do not have Parkinson's, the hope is that the gene-targeted treatments could delay or even prevent the disease's occurrence. Recent research indicates that the LRRK2 protein is inappropriately active in individuals with Parkinson's even if they do not carry the LRRK2 mutation.[41] Consequently, these drugs, if effective, may be beneficial for many people with Parkinson's.[42]

Patti Meese hopes to participate in upcoming clinical trials of LRRK2-targeted therapies. She will keep helping further science, she says, "so that no one ever has to hear those four terrible words again: 'You have Parkinson's disease.'"

LRRK2 is not the only genetic target for new therapies. In 2004, National Institutes of Health researchers discovered mutations in a gene called GBA that can be found in about 5–10% of individuals with Parkinson's.[43] Those who develop the disease at a younger age are more likely to carry them.[44] How the mutations lead to Parkinson's is under investigation.[45] However, these mutations, or changes, in the GBA gene decrease the activity of a protein. This protein is responsible for helping break down alpha-synuclein and other proteins in a part of the cell that acts like a garbage incinerator. The less active the protein that is controlled by the GBA gene, the less well the incinerator works. When this is the case, more garbage bags of alpha-synuclein accumulate in cells. Indeed, Lewy bodies, the garbage bags of alpha-synuclein, are found in the substantia nigra of individuals with GBA mutations.[46]

A major investor in gene-targeted therapies is one of the world's leading venture capitalists—and someone who stands to benefit personally from advances. In 2017, Jonathan Silverstein had just made *Forbes* magazine's "Midas List" of dealmakers for the sixth time when he noticed a tremor in his left leg. He was forty-nine. His subsequent Parkinson's diagnosis was attributed to a GBA mutation.[47]

Silverstein was at first reluctant to disclose his illness publicly. He said, "No one wants to be known as a sick person. . . . Everyone wants to be known as a Super Bowl–winning quarterback." He realized, though, that he had the resources to accelerate efforts to find a cure. "I knew I could put a spotlight on [Parkinson's disease]," he said.

Given his work as a prominent venture capitalist in the biotech and medical devices sectors, Silverstein was well positioned to help advance new treatments for his disease. Companies that he and his firm have backed have received FDA approval for over eighty devices, diagnostics, and therapeutics for a wide range of conditions.

Silverstein started by mailing 1,000 letters to fellow financiers, drug developers, and scientists asking them what were the most promising areas of research and treatment for Parkinson's due to GBA mutations. The responses came back with many ideas. Given his high profile and appearances on CNBC, in *Forbes*, and in numerous other media outlets, Silverstein received unsolicited letters as well.[48] Among them was one from a man in Ohio with GBA-associated Parkinson's disease who sent a check for $18.[49]

Armed with guidance from the people on the cutting edge of Parkinson's research, Silverstein and his wife created the Silverstein Foundation for Parkinson's with GBA. The group aims to educate people about the disease (most people with Parkinson's due to GBA mutations are unaware of the cause) and to develop new therapies to both treat and prevent the disease in those who carry the mutation.[50] "If I don't do it," Silverstein said, "who does?"[51] In its first year, the foundation and its partners launched Prevail Therapeutics, a biotechnology company that has since raised $129 million for the development of gene therapies.[52] In 2018, Silverstein was recognized on the "Midas List" for a seventh time.

Clinical trials of treatments aimed at GBA have recently begun. One of the first drugs to be evaluated is called ambroxol.[53] It could improve the function of the protein that is altered in people who have Parkinson's due to GBA mutations.

In 2017, the biotechnology company Sanofi Genzyme launched a clinical trial of another drug aimed at treating Parkinson's due to the GBA mutation.[54] It was designed to lower the levels of a fatty substance that builds up as result of the GBA mutation.[55] The study, which is seeking to enroll 243 participants, is expected to end in 2022.[56] Prevail Therapeutics is hoping to take up to four drugs into clinical trials in 2019.

Therapies directed at these various mutations mark the beginning of an era of personalized medicine for Parkinson's—an advance that is long overdue. The Parkinson's Foundation is helping prepare for this step toward individualized treatment. Formed in 2016 following a merger of two foundations, the Parkinson's Foundation seeks to improve care and advance research toward a cure.[57] In 2018, the Parkinson's Foundation launched "PD GENEration: Mapping the Future of Parkinson's Disease" to offer genetic testing and genetic counseling to up to 15,000 people diagnosed with Parkinson's disease.[58] While genetic therapies are not yet available, genetic information about what is or is not causing a person's Parkinson's is useful. With appropriate support, people can learn more about their prognosis—genetic causes of Parkinson's have different rates of progression. Genetic testing also opens the door to participating in certain research studies and, if desired, informing family members about their own risk.[59]

EMPOWERING THE IMMUNE SYSTEM TO FIGHT PARKINSON'S

Immunization, the same tool that we use to fight infectious diseases, is now being tested as a possible treatment for Parkinson's (**Box A**). When we are infected with a virus, such as chicken pox or the flu, our immune systems produce antibodies to kill it off. Our immune system also reacts to foreign proteins. This is convenient in the case of Parkinson's because misfolded forms of the alpha-synuclein protein—that major factor in the disease—could be targeted. Unfortunately, the human immune system, for uncertain reasons, does not clear the misfolded protein. Scientists are now experimenting with ways to immunize people with the disease so that the misfolded protein is removed or at least does not spread.

Before polio vaccines were developed to prompt people's immune systems to make their own antibodies, Dr. William Hammon of the University of Pittsburgh had a different idea. He wanted to give individuals existing antibodies against polio to see if they would protect against future infection. Hammon first obtained antibodies from people who had been infected with polio and recovered on their own. He then injected these personalized fighters into children who had not

---

BOX A. WHAT IS IMMUNIZATION?

Any time our immune system comes into contact with foreign viruses, bacteria, or proteins, it produces antibodies against them. These antibodies—proteins that fight infections—then remain on guard, prepared to pounce if the pathogen shows up again. With vaccines, we can trigger this protective process by introducing a weakened or dead form of a given bacterium, virus, or protein. Then, if we are subsequently exposed to the actual pathogen, our antibodies swing into action, wiping out the invaders and sparing us from becoming ill. In another form of immunization, we can give disease-fighting antibodies from one individual to another in order to fight infections.

been infected. The children received modest benefit—fewer of them got polio than those who were not immunized. However, Hammon knew the approach had limitations.[60] The antibodies were in short supply and insufficient to immunize entire communities. The other issue was that the effects tended to wane, so repeated injections were required. When Jonas Salk and Albert Sabin developed effective vaccines, Hammon's treatment fell out of favor.[61]

In October 2018, researchers tried Hammon's approach against Parkinson's.[62] They gave eighty people three monthly infusions of an antibody directed against the toxic misfolded forms of the alpha-synuclein protein. The antibodies did lower the levels of these proteins in the blood and in the fluid surrounding the brain, where it matters most. Many companies are now pursuing this approach for Parkinson's.[63] While promising, the approach requires much more investigation.[64] As with polio, vaccines against these malformed alpha-synuclein proteins is an exciting prospect.

Vaccine studies have recently begun. In 2017, an Austrian biotechnology company called AFFiRiS announced that in an early study of thirty-six people with Parkinson's, its vaccine did not result in serious safety concerns and was tolerated well.[65] The vaccine also resulted in the production of antibodies against alpha-synuclein in about half the participants. However, their antibody levels declined over time and required a later booster injection.[66] These results are encouraging, but much more work remains, including a demonstration that the vaccine reduces symptoms in those with the disease.

MAKING NEW THERAPIES ACCESSIBLE TO ALL

In the United States, prescription drugs account for only 5% of healthcare expenditures on Parkinson's compared to 14% for all conditions.[67] That is largely because levodopa is a generic drug and is cheap compared to most medications. New treatments, as some patients are discovering, are much more costly.

Four years ago, Karen Blair (not her real name), a seventy-year-old architect, was diagnosed with Parkinson's. Her neurologist prescribed a medication called rasagiline (brand name Azilect) that improves the symptoms of the disease modestly.[68] Studies in laboratory animals suggest that, unlike levodopa, the drug also slows disease progression.[69] However, demonstrating such effects in humans is very difficult in the absence of a reliable measurement of its advance.[70] Still, Blair was eager to start the drug. When she did, she got a shock.

Despite having prescription drug coverage through Medicare, the medication cost Blair $600 a month. She tolerated it well and liked the idea of taking something that might slow her disease. But the cost was eating away at her retirement savings. She did not feel much better taking the drug, so she stopped it. In doing so, Blair joined the 45 million Americans who forego medications because they are too expensive.[71]

To be sure, developing safe and effective drugs requires a huge investment. The Tufts Center for the Study of Drug Development and its collaborators estimate that the research and development costs for a new therapy are as high as $2.6 billion.[72] Other estimates are lower but of similar magnitude.[73] Regardless of the exact numbers, drug development is a risky business. Up to 90% of drugs that show benefit in laboratory animals and are then tested in humans fail.[74]

Some drugs have enormous health and economic value. Drugs for HIV changed the infection from a death sentence into a manageable disease. As we know, drugs can cure infectious conditions and even some cancers. In the absence of other treatments, many of these medications justify their high price. The problem is that in the United States, most do not. Instead, pricing is based on what pharmaceutical companies can charge in a very imperfect market.[75] And when their prices are astronomical—justified or not—Medicare and other insurers do not pick up enough of the cost.

The 2019 closure of Medicare's "doughnut hole," a previous gap in the program's coverage of expensive drugs, will help some.[76] The real solution is universal health insurance coupled with caps on total expenses that patients pay. If someone is sick and a highly effective therapy is available, a country that is as wealthy as ours can afford to cover the cost of that therapy.

For high-priced therapies that offer the promise of great benefit, new models are emerging. One, for example, requires insurers to pay only if a drug works (e.g., a cancer therapy cures cancer).[77] Highly effective treatments for Parkinson's should reduce the estimated $25 billion annually that is spent on medical care, mostly by Medicare, for the disease. A cure, of course, would eliminate such expenditures.

To lower overall drug spending, Medicare (paid for by taxpayers) could use its purchasing power to negotiate lower prices, especially for drugs that have high prices but only modest value.[78] In addition, generic drugs should be permitted rapid entry once drug patents expire, and anticompetitive behavior that delays entry should not be tolerated.[79] All these approaches have limitations, but new ideas and more work are needed to ensure that future effective therapies for Parkinson's can help everyone.

EXTENDING LEVODOPA'S REACH

Even current inexpensive medications—the generic levodopa can cost as little as $11 a month—are unattainable for most patients.[80] The main reason is that the drug is not available in all countries. A 2004 report by the World Health Organization found that 40% of countries lacked access to levodopa, still the gold standard treatment.[81] Not surprisingly, access varied by a country's income. While 84% of high-income countries had access to levodopa, only 17% of low-income countries did.

The World Health Organization includes levodopa on its list of "essential drugs."[82] The list is supposed to help countries and their departments of health prioritize their spending on medicines.[83] However, many countries have yet to make purchasing levodopa a priority.

Given limited access to medications for Parkinson's, individuals in some countries are left to utilize alternative sources for levodopa.

FIGURE 2. Photograph of *Mucuna pruriens*, the velvet bean plant.

It turns out that a bean plant called *Mucuna pruriens* is an excellent source of the dopamine precursor. The plant (**Figure 2**) is found in many tropical and subtropical areas where access to medications is limited, and in some countries, such as India, it is used as a treatment for Parkinson's.[84] Clinical trials suggest that *Mucuna pruriens* may have comparable benefits to medical preparations of levodopa.[85] However, the plant is not available everywhere levodopa is needed. Over-the-counter formulations are available in the United States, but dosing and formulations are widely variable. In addition, levodopa alone can cause nausea and other side effects. In most medical formulations, levodopa is combined with another drug to reduce these side effects. Fifty years after its development, we need to ensure that the miraculous levodopa is available to all.

9

TAKING CHARGE

The Policies and Research Funding We Need

Our lives begin to end the day we become silent
about things that matter.

—Martin Luther King Jr.

ON FEBRUARY 22, 2006, AT THE WALTER E. WASHINGTON CON-vention Center in Washington, DC, 3,200 people gathered for the World Parkinson Congress, the first major event to bring researchers and people with the disease together. The speakers included leaders from the National Institutes of Health (NIH), leading clinicians, and top scientists. And Tom Isaacs.

Eleven years before, Isaacs had been a twenty-six-year-old sur-veyor working for a London real estate company when he developed a tremor and was subsequently diagnosed with Parkinson's. When he told his parents, he said, "I think I could be a good bartender. I have a really good shake."

From the start, Isaacs was focused on a cure. His goal was to be among the first people to be able to say, "I used to have Parkinson's." At that time, he said, "the word cure was never used. . . . [Y]ou knew it was forbidden. Cure was seen as a false hope. And actually if you don't have hope in Parkinson's disease, you don't have anything."[1]

Isaacs left London on April 11, 2002, to start a different journey. Over the next year, he walked 4,500 miles around the coast of Great Britain—"anticlockwise," as he put it. He climbed the highest mountains in England, Scotland, and Wales, wore through five pairs of boots, stayed at 238 different bed-and-breakfasts, walked with 1,032 different people, and ultimately raised $500,000 for Parkinson's research (**Figure 1**). Days after finishing the journey, he ran the London Marathon.[2]

In 2005, Isaacs founded the Cure Parkinson's Trust with three friends.[3] They wanted "something focused on the research . . . and it needed to be a bit more edgy, a bit more feisty."[4] They did not just

FIGURE 1. Tom Isaacs crossing the Millennium Bridge after walking around the perimeter of Great Britain, 2002. *Courtesy of Lyndsey Isaacs.*

want to fund researchers and then wait for results. They wanted to drive progress.

Isaacs was relentless in his pursuit of a cure. He raised funds, met with countless scientists, visited Pope Francis to discuss the potential of stem cells (the Catholic Church opposes the use of stem cells derived from embryos but not other sources, such as adult stem cells), and volunteered to participate in clinical trials of experimental drugs.[5] He was humorous, self-deprecating, and optimistic.

Isaacs believed that new treatments and a cure were only possible through the active participation of those affected by the disease. He said,

> There is one thing that motivates me more than anything; it is the idea of finding a deliverable treatment that can reverse the course of this cruel illness. I believe that this is attainable. For me, one of the biggest challenges—and yet also the biggest opportunity—is galvanizing people who live with Parkinson's from day to day to engage with their condition. If everyone with Parkinson's were to communicate their experiences of living with Parkinson's; if everyone participated in clinical trials; if everyone took the time to become more knowledgeable and was more committed to partnering [with] the scientific community in the search for new treatments, then there is no doubt in my mind that progress can be accelerated in Parkinson's research.[6]

He established this partnership at his foundation, making sure that every research project it supported was evaluated by and relevant to people who had Parkinson's. Though Isaacs died in 2017 at the age of forty-nine, the work at the trust continues as he intended—relentlessly, collaboratively, and joyfully.[7]

CHANGING POLICY TO PREVENT THE DISEASE

We can prevent many cases of Parkinson's. Paraquat and trichloroethylene (TCE) are two of the biggest known culprits contributing to the rise in Parkinson's, and the Environmental Protection Agency (EPA) has both the power and the responsibility to remove these threats. If it does, fewer of us will develop Parkinson's. It is that simple.

In addition to the over 100,000 individuals who have signed a petition asking the EPA to ban paraquat, Marine veterans and their families are also demanding that the EPA ban TCE.[8] They and their families have suffered terrible health consequences from the chemical and routinely go to Capitol Hill to pressure the EPA. Retired Master Sergeant Jerry Ensminger, who lost his daughter to leukemia at Camp Lejeune, said at a 2018 press conference, "What the hell are we waiting for? What is the EPA waiting for?"[9] The EPA is failing to fulfill its core mission of "protecting human health and the environment [and of] providing clean and safe air, water, and land for all Americans."[10]

In other parts of the government, there has been some movement. In 2016, as part of the 21st Century Cures Act, Congress authorized a National Neurological Conditions Surveillance System. In 2018, President Donald Trump signed a bill to fund the system with $5 million. The monies will initially be used to track and collect data on the number of people with Parkinson's and multiple sclerosis.[11] These efforts should help scientists identify geographic clusters of disease, evaluate potential risk factors, and analyze disease trends. While $5 million is a start, additional funds will be needed to support identification and assessments of risk going forward.

Looking for other risk factors is important, but it should not delay action on what we already know will lower all of our risk of developing Parkinson's.

TURNING FRUSTRATION INTO ACTION

Dr. Andy Grove, the co-founder of Intel, understood that action had to be forceful. He dedicated much of the last years of his life to fighting Parkinson's following his diagnosis in 2000.[12]

Grove was born to a Jewish family in Budapest, Hungary, in 1936. When he was eight, the Nazis occupied Hungary and deported half a million Jews to concentration camps. With a fake name and assistance from friends, Grove's family escaped detection.[13] Later, following the Hungarian Revolution of 1956, he and his family fled to Austria. They eventually made their way to the United States.

Six years later, Grove joined the pioneering chip manufacturer Fairchild Semiconductor. He later left to help form Intel, a company that he would eventually lead. Grove would transform the chip maker into an undisputed leader in Silicon Valley and become a respected thinker in business management circles.[14]

After Grove was diagnosed with Parkinson's, he became frustrated by the lack of progress in terms of finding a cure. He said, "You can't go close to this and not get angry. There are so many people working so hard and achieving so little."[15] True to the Silicon Valley way, Grove sought to disrupt the status quo. In a speech to several hundred of the world's leading scientists, he said, "What is needed is a cultural revolution that values curiosity, follow-through and a problem-solving orientation." He called on the scientists to put their data "in full view, scrutinizable by all." By sharing data, researchers could minimize repeating experiments, avoid fruitless investigations, and build on the insights of other researchers.

He wrote to the NIH's director about how to improve Parkinson's disease research. He received no response, but that did not slow Grove.[16] Unable to make changes at NIH, he made his own. With the University of California, he created new educational programs that married engineering principles with biological sciences.[17] He pushed for objective measurement of Parkinson's symptoms with new devices and wearable sensors.[18] He insisted on data-sharing platforms. He gave millions to research.[19] Grove was, according to the *Washington Post*, "an advocate and agitator for better and faster medical research."[20] His efforts have accelerated research cycles, broadened the sharing of research data, and spurred the use of technology in the Parkinson's field.[21]

Many scientists benefitted from Grove's mentorship. His style was blunt and direct.[22] He guided hundreds and asked his protégés what their career goals were. If they were not clear enough, he would tell them to give it more thought. He told the cofounder of the Michael J. Fox Foundation, Deborah Brooks, "You have an obligation to behave exponentially different than everybody else."[23] Dr. Jeffrey Kordower, a leading Parkinson's disease researcher at Rush University, said that Andy Grove "was incredibly direct and forced me to focus on problem solving and stay central to the core of what we were studying. He was an absolute hero. He changed the world."[24]

Grove helped create Silicon Valley, developed new management principles, advanced cancer research (he survived prostate cancer), and provided educational scholarships for thousands around the world so that they can have the educational opportunities that he did.[25] Upon his death, Grove dedicated $40 million of his estate to Parkinson's research in partnership with The Michael J. Fox Foundation. Even without his resources, Grove's dissatisfaction with the status quo drove change, an achievement that we can all strive for.

CLOSING THE FUNDING GAP IN PARKINSON'S RESEARCH

The NIH is the largest public funder of biomedical research in the world, spending $37 billion in 2018.[26] However, its support for Parkinson's research has waned over the past decade, just as the numbers of people with the disease have shot up (**Figure 2**). Adjusted for inflation, the NIH spent $192 million on Parkinson's disease research in 2008. Ten years later, in 2018, it spent $177 million, or 8% less.[27] At the same time, the estimated number of Americans with Parkinson's increased 40%.[28] Without advocacy, even well-intentioned organizations like the NIH can move in the wrong direction despite the tremendous increase in the burden of Parkinson's disease.

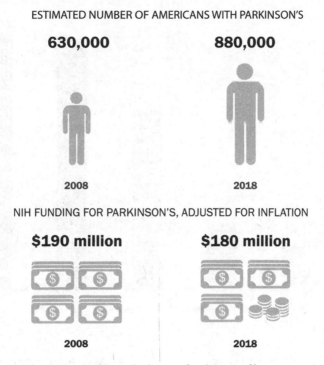

ESTIMATED NUMBER OF AMERICANS WITH PARKINSON'S

630,000   **880,000**

2008   2018

NIH FUNDING FOR PARKINSON'S, ADJUSTED FOR INFLATION

$190 million   **$180 million**

2008   2018

FIGURE 2. Parkinson's disease funding gap.[29]

The decline in US public support for Parkinson's research has had an unlikely savior—a Canadian. In 1991, Michael J. Fox was a thirty-year-old film and TV star. While filming *Doc Hollywood*, he noticed a tremor in his left hand and that his arm wasn't swinging normally when he walked; he was eventually diagnosed with Parkinson's.[30]

Nine years after his diagnosis, he created The Michael J. Fox Foundation for Parkinson's Research. Two years before that, in testimony to Congress, Fox said that he "was shocked and frustrated to learn the amount of funding for Parkinson's research is so meager."[31] Since its establishment, the foundation has devoted $900 million to research programs, becoming the largest nonprofit funder of Parkinson's

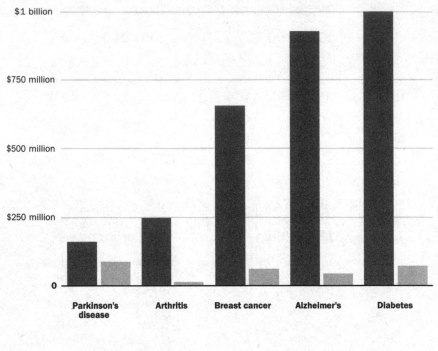

FIGURE 3. Comparison of NIH funding with research expenditures by the largest charitable foundations for select diseases, 2016.[32]

research in the world. The organization has no endowment. It spends all the money that it raises in a given year. In 2018, it spent nearly $100 million on research (**Figure 3**).

The foundation invests in studies and builds tools, such as genetic databases, for the research community to use. Its research portfolio spans basic research with rodents to clinical research involving humans. The foundation also funds therapeutic projects in partnership with universities and pharmaceutical companies, and its early-stage investment propelled inhaled levodopa to market in 2019.[33] By bypassing the digestive system, the inhaled levodopa can provide faster relief of symptoms.[34]

UNDERSTANDING THE CAUSES OF PARKINSON'S DISEASE

The Michael J. Fox Foundation has a worthy goal: to cure Parkinson's disease. This is the common hope for many groups fighting various illnesses. However, the greatest opportunity may be in preventing Parkinson's in the first place.[35]

Dr. Caroline Tanner is the neurologist at the University of California, San Francisco, who has led numerous studies that have linked pesticides, TCE, and other environmental factors to Parkinson's.[36] Despite this groundbreaking work, she has not been able to do further large-scale studies aimed at identifying environmental risks. Why? "No one was willing to fund the research," she said. Some efforts are beginning, but more are needed.[37]

To better understand environmental causes of Parkinson's, simple maps that show the rates of the disease by geography would be a great start. These maps of "hot spots" overlaid with information about pesticide use, TCE contamination, air pollution, or other environmental factors could help scientists and others identify what might be causing the disease.

Researchers have maps of hot spots for cancer. Through a public NIH website, anyone—citizen or scientist—can find the rates of any cancer for any part of the country.[38] Anyone can see where the rates of colon cancer, for example, are highest or lowest and what the risk is by age or race. We can also see what the trends are over the past five years.

For Parkinson's there is very little of this helpful information. Studies that identify geographic variability, such as those showing that people who live in some rural areas have higher rates of the disease, make a strong case that environmental factors lead to the development of Parkinson's.[39] What we lack are details and real-time data. We cannot zoom in on any specific area and determine what the rates of Parkinson's are or learn what pesticides or other chemicals have been or currently are being used. With rare exceptions, neither citizens nor researchers can determine the trends in the rates of Parkinson's disease in their states.[40]

For a decade, Tanner and her colleagues have worked to develop such a map for California through a statewide registry of Parkinson's.[41] On July 1, 2018, the California Parkinson's Disease Registry was finally launched.[42] Previously, Nebraska and Utah were the only states to have such a registry.[43] Data from the Nebraska registry allowed scientists to see that the rural, farming regions had two to four times more Parkinson's than urban areas and that men had roughly twice the rates of the disease as women.[44] Additional studies should quantify the risk posed by specific environmental risk factors and correlate those exposures with subsequent rates of the disease. And, of course, we need to be looking at every state—not just three.

Studies of Parkinson's need to be large. Even though the number of people with the disease is rising, the total number is small relative to very common conditions such as diabetes. So in order to get reliable data and see trends clearly, large populations have to be studied for years. Such studies could track the rates of Parkinson's among people living near TCE-contaminated Superfund sites. They could allow researchers to determine if certain groups of individuals, such as dry cleaners, are more likely to develop Parkinson's.

These efforts should also be expanded globally. For example, identifying potential risk factors for the rapid rise of Parkinson's disease in China is critical to slowing the disease.[45] The risks from pesticides and air pollution could be quantified and compared to one another. Such data could inform what environmental actions should be taken and what short-term protections should be offered.

The genetic causes of Parkinson's are also worth further investigation. Over the past two decades, Dr. Andy Singleton, who is now chief of the Laboratory of Neurogenetics at the NIH, has led pioneering genetic studies of the disease. Six years after researchers identified the first mutation in the alpha-synuclein gene, he and his colleagues identified another rare mutation that causes an early-onset, inherited form of Parkinson's disease.[46] One year later, he and his research team identified mutations in the LRRK2 gene that lead to Parkinson's.[47]

Before the genetic causes of the disease were identified, few good scientific targets for new therapies were available to researchers. Now, based on the work of Singleton and Dr. Matthew Farrer from the University of Florida and Dr. John Hardy from University College London, and others, drug developers have something to go after, which represents a huge leap forward toward a therapy or a cure.[48]

Again, cancer offers a model. Treatment aimed at the underlying genetic causes of the disease is transforming care. Cancer treatments are increasingly based on the individual characteristics and often underlying genetic mutations responsible for the tumor. For example, in melanoma, one class of drugs is effective if the skin cancer has a certain genetic mutation. But in those cancers without the mutation, the drugs are unlikely to work and can cause harmful side effects.[49] This personalized approach, where a drug is aimed at the underlying genetics of the disease, now applies to colon cancer, leukemia, lung cancer, and prostate cancer too.[50]

Now that we have identified relevant genetic causes, the same could happen for Parkinson's.[51] Armed with greater resources, scientists can identify more of these targets and learn how to neutralize them, as is already happening with the drugs in development to defang the threat of an LRRK2 or GBA mutation.

In 1991, Hyam Kramer's mother was diagnosed with Parkinson's disease. Kramer asked her neurologist if the disease could be inherited, and he was told no. Twenty-five years later, Kramer developed a tremor in his right hand and right leg and was subsequently diagnosed too. What his mother's neurologist could not have known was that a mutation in the GBA gene was responsible for her Parkinson's.[52]

Kramer's mother, a former librarian, lived with Parkinson's disease for twenty-two years. She was cared for by her husband, a former World War II veteran, who would visit her daily in the nursing home where she lived for the last twelve years of her life. She required a feeding tube, and at the end of her life, she could no longer recognize her son.

Watching a loved one decline is difficult enough, but Kramer was also (unknowingly) witnessing his own future. Normally optimistic and confident, Kramer, a professional fund-raiser, became depressed. Still, he signed up for a Michael J. Fox Foundation–funded study called the Parkinson's Progression Markers Initiative.

To help identify objective measures of disease progression, The Michael J. Fox Foundation launched the study in 2010. This global initiative has enrolled nearly 1,400 participants at more than thirty research sites. Every six to twelve months, participants undergo detailed clinical assessments, imaging tests, and sampling of their blood and the spinal fluid surrounding their brain and spinal cord. The goal is to identify biological markers (akin to cholesterol for heart disease) that will track the progression of Parkinson's disease.[53] The data—available to scientists around the world—have been downloaded over 5 million times by research teams from over thirty countries.[54] Unfortunately, while knowledge has advanced tremendously, no objective measure of disease progression has yet been identified.[55]

For the study, researchers were looking for Ashkenazi Jews with Parkinson's, and Kramer fit the bill. He learned that he carried mutations in both copies of his GBA gene. (Humans have two copies of every gene—one from their mothers, one from their fathers.)

As a participant in the study, Kramer, who lives in Boston, Massachusetts, travels every six to twelve months to New Haven, Connecticut, to meet with researchers. During these visits, the scientists take various measurements so they can understand the course of Kramer's disease. They are looking at markers in his blood and spinal fluid and are using sophisticated imaging tests to see how the disease progresses. As of April 2019, seventy-three individuals like Kramer with Parkinson's disease due to a GBA mutation had enrolled in the global study.

Kramer has already participated in a clinical trial of a new drug and intends to participate in more. He hopes that his involvement in research studies and clinical trials will "shed light onto future treatments that [he] or someone like [him] can take advantage of."

For Kramer, there has been a side benefit too. Helping the science move forward has brought back his optimism. Now fifty-eight, he recently married his longtime partner. He says that participating in the research has given him a sense of making a positive impact. "In some ways, the last three years have been the best three years of my life," he says.

UNDERSTANDING HOW PARKINSON'S DEVELOPS

Identifying the environmental and genetic causes of Parkinson's is an important first step. The next step is to determine how these factors lead to disease. Animal studies can help us determine how risk factors contribute to Parkinson's. For example, in rats, TCE kills off dopamine-producing nerve cells in the substantia nigra region of the brain.[56] Other parts of the rats' brain are less damaged.[57] But it is unclear exactly how TCE causes this apparent selective loss of nerve cells.[58]

When it comes to how genetic mutations cause Parkinson's, we do not have the whole story either. We know that the alpha-synuclein protein helps transport packages of neurotransmitters, enabling communication among nerve cells. We also know that these proteins can be misfolded—likely as the result of a genetic mutation—which leads to Parkinson's.[59] However, the steps from misfolding to disease remain to be identified. The functions of the alpha-synuclein protein, including those outside the brain, have also not been defined.[60]

In addition, we do not have a full understanding of the LRRK2 gene and all its mutations.[61] Researchers are still determining the various functions of the protein that the LRRK2 encodes, or instructs. The protein that takes its orders from the LRRK2 gene is large and likely interacts with the alpha-synuclein protein.[62] The changes in the protein that result from the different mutations are even more of a mystery.

Basic research, with its exploratory nature, could help fill in some of these gaps. Unlike applied research, basic research has no predefined purpose. Rather, it helps us understand how nature works. Unfortunately, it is often overlooked, even though therapeutic advances rest on knowing how a disease—a change of nature—unfolds.

Over the last twenty-five years, pharmaceutical companies have invested less and less in basic research.[63] In 1994, pharmaceutical companies spent over half (57%) of their research and development funds on this early-stage research.[64] By 2017, that percentage had dropped to 16%.[65] Consequently, the responsibility for this work has fallen to academic labs and research institutes that are heavily dependent upon federal support and private philanthropy.[66] In the United States, most biomedical research (about 60%) is funded by pharmaceutical, biotechnology, and medical device firms. The upshot is that basic research is dependent on smaller sources of funding—the NIH and private foundations, which represent 27% and 4%, respectively, of the total research pie.[67]

Increasing our knowledge of the function of different genes is critical to understanding how the interaction of genetic and environmental factors can lead to Parkinson's disease.[68] Dr. Tim Greenamyre, who helped link pesticides to Parkinson's, and his colleague Dr. Jason Cannon recently wrote, "Many researchers believe that gene-environmental interactions account for the majority of Parkinson disease cases."[69] They called for "[future] studies examining as many known genetic factors as possible."[70] These studies could inform us about how genetic factors influence the response to pesticides, chemicals, and other environmental risk factors.[71]

Finally, more research is needed to determine how various factors, including caffeine, diet, and exercise, decrease the risk of Parkinson's. This information is critical because it can empower us to take advantage of behaviors that are available to most of us. Eating well and exercising carry many secondary health benefits, of course, and have been overlooked for too long.

DEVELOPING BETTER MEASURES OF PARKINSON'S DISEASE

The management guru Peter Drucker once said, "You can't manage what you can't measure." Unfortunately for Parkinson's, we don't measure the disease well.

For heart disease, we have electrocardiograms, echocardiograms, angiograms, blood cholesterol levels, and blood pressure. For cancer, we have blood markers, imaging, and biopsies. For HIV, we can quantify the amount of virus in the blood and count the number of blood cells infected by the virus. For Parkinson's, the principal tools of the twenty-first century are the same ones that Drs. James Parkinson, Jean-Martin Charcot, and William Gowers used in the nineteenth century—a patient's history and a physical examination. In other words, we have almost no objective measures of the disease.

The history and exam can be misleading even in the best hands. Autopsies of people with diagnosed Parkinson's reveal that their doctors had been wrong about their diagnoses 10% to 20% of the time.[72] People thought to have Parkinson's disease were shown to have had different parkinsonian disorders. Others had Alzheimer's disease or tremors due to another cause.[73] Unfortunately, over the last twenty-five years, doctors have not gotten better at diagnosing the disease.[74]

Just as diagnosis of the disease remains subjective and error prone, so too does the measurement of symptoms. Currently, the determination of the severity of Parkinson's disease symptoms, such as difficulty walking, is based on clinical observations by experts. The most commonly used Parkinson's measure is a rating scale that includes a "motor examination" by a clinician.[75] The examination requires a neurologist or other trained investigator to rate multiple features of Parkinson's disease, such as tremors and speed of movement, on a simple but crude 0-to-4 scale.

Not surprisingly, experts can vary in their assessment of how pronounced the movements are. One may rate a tremor a 2 (mild) while another may rate it a 1 (slight). Also, because the rating scale uses a

0-to-4 scale, it is insensitive to small changes in the disease. Small changes can matter over a short period when you want to know if a treatment is making someone better or worse. For example, a man's weight can go from 184.8 pounds to 181.4 pounds—enough to see if a weight-loss program is really working. However, his tremor cannot go from 3.4 to 2.9—only from 3 to 2. So, with existing scales, it is easy to miss genuine but small improvements in Parkinson's symptoms. Dr. Andy Grove, the activist engineer, referred to a Parkinson's disease rating scale as "a piece of crap."[76]

When traditional approaches are not working, sometimes voices from outside the field can offer more creative approaches and fresh hope. As a kid, Max Little was a mathematical whiz who, like many young boys, enjoyed video games. His twin passions fused when he developed digital sound algorithms for video games, which enable players to hear unique sounds like a dog barking or cars racing. He then earned a PhD in applied mathematics at the University of Oxford in England, where he created new mathematical rules and techniques for measuring voice.[77]

He and his colleagues then used these techniques to assess the voice of individuals with Parkinson's, which is softened early in the disease.[78] In 2012, the researchers showed that mathematical analysis of voice recordings could tell who had Parkinson's and who did not.[79] The results could even be used to predict disease severity.[80]

In 2012, Little gave a TED talk titled "A Test for Parkinson's with a Phone Call."[81] In it, he explained how his test could be administered with any digital microphone. He also used the talk to launch a worldwide study, the Parkinson's Voice Initiative, which, within hours, had recruited over 1,000 participants to leave recordings of their voice over the telephone. After just a few weeks, it had collected over 17,000 voices and remains the largest "citizen-led" scientific study of Parkinson's disease.

Little and his colleagues went on to develop a smartphone application that can take additional objective measurements of symptoms. Using the phone's embedded sensors, the app measures gait, balance, finger tapping (to assess speed of movements), and voice.[82] In a pilot study, the app could differentiate individuals with Parkinson's from those without the disease.[83]

In subsequent research, those same assessments derived from smartphone sensors generated a "mobile Parkinson disease score." This score combined assessments of voice, tapping, and walking into a single continuous measure of disease severity. The scale goes from 0 (normal) to 100 (severe disease) and can be performed by almost anyone with a smartphone anywhere anytime.[84]

Inspired by the research of Little and others, Apple Inc. developed an open-source platform called ResearchKit for smartphone applications. The platform enables anyone to build research studies on smartphones for any condition from autism to Alzheimer's disease. On March 9, 2015, Jeff Williams, the current chief operating officer of Apple, announced to a packed audience and about 1 million people watching online the launch of five smartphone research studies, including one for Parkinson's disease called mPower **(Figure 4)**.[85]

In his talk, Williams identified several barriers to medical research—participation restricted to people who live near research

FIGURE 4. mPower smartphone application, 2015.

sites, subjective data, and infrequent assessments—all of which the mPower app eliminates. With it, individuals can use their iPhones to enroll in the study, conduct objective assessments frequently, and receive their results back in real time.[86] They can see their individual voice or tapping scores as soon as they complete the tests. In most studies participants only learn the results after the research is completed. Some never learn them at all.[87]

Williams gave his talk in the morning. By dinner, more than 2,000 individuals from across the country had enrolled in the Parkinson's study. By contrast, the largest Parkinson's disease clinical trial of a drug took three years to enroll 1,741 participants.[88] Combining the Parkinson's Voice Initiative's approach of citizen-led recruitment with smartphones created a new paradigm for research studies.[89]

The science continues to accelerate. In 2017, a smartwatch study was launched to screen for abnormal heart rhythms and enrolled over 400,000 people in one year.[90] In 2018, Apple publicly announced that the Apple Watch could also track symptoms of Parkinson's disease.[91] That same year, Verily Life Sciences (part of Alphabet, the parent company of Google) partnered with Radboud University Medical Center in the Netherlands and with The Michael J. Fox Foundation to investigate the ability of a watch to assess tremors, activity, and sleep.[92] The initial results on adoption are encouraging— hundreds of Dutch patients are wearing the Verily smartwatch for twenty-two hours per day; some have done so for more than a year. Such observations allow insights into how people function in their own environments, which can differ from their performance in artificial clinical settings.

Objective measurement of the effects of new therapies is critical. Current subjective rating scales can lead to ambiguous results. For example, two separate studies of the same drug for Parkinson's that were conducted with a nearly identical design and outcome measures had completely different results.[93] Imprecise measurements have real consequences—they make it difficult to determine whether a new treatment works for Parkinson's. As a result, clinical trials for the disease and similar disorders end up having to be larger, longer, more expensive, and riskier than for other conditions.[94]

Because of the risk and cost of clinical trials in conditions like Parkinson's, large pharmaceutical companies may be reluctant to invest in them. On January 8, 2018, Pfizer announced that it was ending research into new drugs for Alzheimer's and Parkinson's.[95] High-profile failures to find an effective drug for Alzheimer's and the absence of substantial therapeutic progress in Parkinson's disease likely contributed to the decision.[96]

Smaller companies are filling the void left by larger companies. In fact, about one year after Pfizer's exit, it formed a spin-off company called Cerevel Therapeutics. Backed by $350 million in venture funding, Cerevel Therapeutics will develop new neurological drugs, including a promising one for Parkinson's disease that acts on dopamine.[97] The new company hopes in 2019 to begin a late-stage clinical trial of the drug, which may not have levodopa's side effects.[98]

Many pharmaceutical companies are embracing digital measures for Parkinson's. Roche used a smartphone application in an early-stage trial for Parkinson's and found it to be more sensitive than traditional rating scales.[99] It is now in use in a subsequent clinical trial. Other pharmaceutical companies are evaluating watches and other sensors with the plan to use them to assess future treatments.[100]

Cerevel and others will need these and other measures to gain an accurate assessment of whether their drugs work. As Little said, "The only way we're going to know when we actually have a cure is when we have an objective measure that can answer that for sure."[101]

OVERCOMING APATHY

In a cruel irony, helping end Parkinson's depends on what the disease often steals—energy and a drive to make things happen. Apathy—a loss of initiative and blunted emotions—is often part of the disease.[102]

The late Leonore Gordon, a former family therapist and resident-poet in the New York City Public Schools, struggled with apathy during her twenty-year fight with Parkinson's.[103] But Gordon ultimately won.

She fought with activism. She figured out that helping others reduced her feelings of indifference. She gave talks, wrote poems, raised funds, and counseled hundreds. To help build awareness of how Parkinson's could pull people down, Gordon participated in an educational television episode on apathy for the Parkinson's disease community. On the show, she said that when people with the disease "get involved in advocacy or volunteering or being involved with anything where they feel needed, we get out of the house." She also said, snapping her fingers, "When I get on the computer and see that someone needs me for help, apathy is gone like that."[104]

Apathy and complacency are dangerous—beyond the dispiritedness that they bring. Our lifetime risk of dying in a car accident is one in one hundred. Consequently, almost all of us wear seat belts, most of us drive cars with air bags, and all of us want children to drive safely.[105] Our lifetime risk of developing Parkinson's is much greater—one in fifteen.[106] And yet, most of us do little.

Like Gordon, we need to overcome our apathy. Unfortunately, Gordon died in 2018. In her memory, her friends wrote, "She was a warrior whose heart was full of love, who had the courage to speak out, to take action, and to motivate others to join her for worthy causes."[107] Let's continue her efforts.

PART THREE

A Prescription for Action

WITHIN OUR REACH

How We Can End Parkinson's

A demand for scientific proof is always a formula for inaction and delay and usually the first reaction of the guilty.... In fact scientific proof has never been, is not, and should not be the basis for political and legal action.

—Scientist at British American Tobacco[1]

WE HAVE TO TAKE ACTION AGAINST PARKINSON'S NOW— through prevention, advocacy, care, and treatment. We end this book with twenty-five concrete steps we can and should take to reduce the worldwide toll of this daunting disease.

PREVENT THE DISEASE

1. Ban paraquat and other harmful pesticides.

According to the Environmental Protection Agency (EPA), the pesticide paraquat is "highly toxic," it has no antidote, and "one sip [of it] can kill."[2] Exposure to it also doubles the risk of developing Parkinson's.[3] It kills dopamine-producing brain cells and produces features of Parkinson's in laboratory animals. Thirty-two countries have already banned paraquat; yet it is still used on crops in much of the United States. In the last decade, its use has doubled.[4]

We have a window to outlaw it now. To be used in the United States, pesticides must be registered with the EPA, and paraquat's registration is up for review. According to the EPA, the federal statute that governs the use of pesticides "prohibits registration of pesticides that generally pose unreasonable risks to people, including agricultural workers, or the environment."[5]

Please contact the administrator of the EPA, currently Andrew Wheeler, and urge him to act against paraquat and its unreasonable risk, including for Parkinson's. You can phone his office at 202-564-4700 or email him at wheeler.andrew@epa.gov. You can also contact the Congressional committees that provide oversight of the EPA and ask them why the United States, unlike dozens of other countries, has failed to ban paraquat. You can contact the US House Committee on Energy and Commerce (202-225-2927) and the US Senate Committee on Environment and Public Works (majority contact: 202-224-6176; minority contact: 202-224-8832).

In the absence of action by the EPA to date, Congresswoman Nydia Velázquez of New York introduced a bill into the US House of Representatives on July 17, 2019, to cancel the registration of all uses of paraquat.[6] To learn about the current state of the bill, go to **advocate .michaeljfox.org**.

In addition to paraquat, several other pesticides increase the risk of Parkinson's disease. Among them is the nerve toxin chlorpyrifos.[7] And yet, according to the EPA, it is the most widely used conventional insecticide in the country.[8] In 2013, chlorpyrifos was used to treat almost sixty different crops covering approximately 1.3 million acres according to the California Farm Bureau Federation, which argues that the pesticide is critical to the local agriculture.[9] Alfalfa, almonds, cotton, grapes, oranges, and walnuts are all sprayed with chlorpyrifos.[10] The pesticide is also used on broccoli, Brussels sprouts, cauliflower, cranberries, soybeans, and Washington State apples.[11]

Golf courses, lawns, utility poles, and wood fences are soaked in chlorpyrifos too.[12] A 2013 letter to the journal *Annals of Neurology* asked, "Is living downwind of a golf course a risk factor for parkinsonism?"[13] The authors reported that nineteen of the twenty-six people that they had seen with parkinsonism lived within two miles of a golf course. In addition, sixteen of the nineteen lived downwind. No conclusion was drawn, but a larger study was sought.[14]

The EPA has considered an outright ban on the chemical since 2007, but in April 2017, the EPA ruled that such a ban was not warranted due to uncertainty about its effects on the brain development of children.[15] In August 2018, a federal court ordered the EPA "to revoke all tolerances and cancel all registrations for chlorpyrifos within 60 days."[16] But one month later the Trump Administration appealed the decision.[17] In the absence of a federal response, California announced in May 2019 that it would ban use of the pesticide.[18] Two months later, the EPA ruled that it would not ban the chemical, saying that the data about its health concerns were "not sufficiently valid, complete or reliable."[19] In January 2019, Congresswoman Nydia Velazquez introduced the Ban Toxic Pesticides Act of 2019 that would eliminate the use of chlorpyrifos in the U.S.[20]

Parkinson's disease may be the least of the risks of chlorpyrifos. A 2006 study in the journal *Pediatrics* found that children born between 1998 and 2002 with the highest prenatal exposure to the bug killer scored lower on development tests than those with lower exposure.[21] Before EPA's 2000 ban on the indoor use of the pesticide, chlorpyrifos was widely used to kill cockroaches, and exposure was widespread among pregnant women in New York City. The ban did not go into effect until the end of 2001.[22] One review calculated that exposure to chlorpyrifos and its class of pesticides among 25.5 million US children led to a total loss of 16.9 million IQ points.[23]

As with most pesticides, the use of chlorpyrifos is not limited to the United States but extends to seventy other countries and "has led to ubiquitous human exposure."[24] Pesticide use is now a huge issue in sub-Saharan Africa, where use has historically been low.[25]

2. Ban trichloroethylene.

Trichloroethylene (TCE) is commonly used as a solvent to remove grease from metal. People can be exposed to it by breathing in its fumes, ingesting it, or absorbing it through their skin. For more than eighty years, we have known about its toxicity.[26] Like paraquat, it can produce features of Parkinson's disease in laboratory animals. In 1977, the Food and Drug Administration (FDA) banned its use as an anesthetic. In December 2016 and January 2017, the EPA proposed banning the use of TCE for industrial degreasing and for spot-cleaning by dry cleaners.[27] In addition, it has identified safer alternatives.[28] Unfortunately, the leadership at the EPA has postponed the ban indefinitely.

In the absence of appropriate regulatory action, legislatures at the state and federal levels are beginning to take action.[29] A bipartisan group of state legislators in Minnesota, led by Republican state senator Roger Chamberlain and Democrat state representative Ami Wazlawik, is seeking to have Minnesota become the first state to ban the chemical.[30] In January 2019, the Minnesota Pollution Control Agency learned that a battery manufacturer outside Minneapolis was emitting much more TCE into the air than allowed.[31] Neighborhoods and residents up to 1.5 miles from the facility were exposed to unsafe levels.[32] One resident testified, "I've lived in my home, one quarter mile from this plant, for 18 years and raised my three children at that home. [The ban is] too late for our community, but we're asking that you ensure this doesn't happen to others in the future."[33]

At the federal level, a serious bipartisan effort is underway in Congress to ban toxic chemicals like TCE. Representative Paul Tonko, Democrat from New York, chairman of the House Energy and Commerce Committee's Subcommittee on the Environment and Climate Change, and Representative John Shimkus, Republican from Illinois, the ranking House Republican, are planning to press the EPA in open hearings in 2019 and 2020. People can attend the hearings, which are open to the public, to show a loud and unified voice on the issue of banning TCE and other chemicals. Banning chemicals is important not only for Parkinson's but potentially for other diseases such as autism.[34] Combining efforts among advocacy groups, which has already begun, can often bring quicker action. The schedule of hearings is available here: **https://energycommerce.house.gov/committee-activity/hearings**.

The Environmental Defense Fund has organized a direct link to email US Senators and Representatives asking them to ban the chemical. You can lend your name to the fight by going to **https://membership.onlineaction.org/site/Advocacy?cmd=display&page=UserAction&id=3254**.

3. Accelerate the cleanup of contaminated sites.

As of July 2018, the United States had 1,346 sites on its National Priorities List—areas with known or threatened contamination with hazardous substances.[35] Of these sites, nearly half are polluted with TCE.

We know where the contamination is, but we are not doing enough about it—at least not at a pace to protect people.

One issue is money. Funding for cleaning up Superfund sites dried up in 1995 when taxes on crude oil, chemicals, and corporations expired.[36] Now most of it comes from taxpayers.[37] To get the program running again and to protect communities surrounding these sites, we need polluting parties to pay and the EPA to assemble teams to rapidly clean the sites.

Superfund sites are only the beginning. The number of TCE-contaminated areas that are not Superfund sites is in the thousands. Michigan alone has three hundred.[38]

In 1974, Congress enacted the Safe Drinking Water Act to ensure the quality of Americans' drinking water, but this does not protect everyone.[39] The water standards only apply to public water systems and not to private wells.[40] Over 40 million Americans get their water from wells. In 2009, the US Geological Survey studied over 2,000 private wells and found that 23% had at least one contaminant of potential health concern.[41] Even public systems can be contaminated. The non-profit Environmental Working Group offers maps (**www.ewg.org**

/interactive-maps/2018-tce) of the systems in the United States that were contaminated with TCE as of 2015.[42]

Those who drink well water should have it tested. The EPA provides guidance on how to do this on its website (**www.epa.gov /privatewells/protect-your-homes-water#welltestanchor**). The EPA recommends using state-certified laboratories or your local health department to do drinking-water tests.[43] More information is available on the EPA's website at **www.epa.gov/sites/production/files/2015-11 /documents/2005_09_14_faq_fs_homewatertesting.pdf**.

TCE also evaporates and can contaminate indoor air without warning. If you live near a site contaminated with TCE, you may want to have your indoor air checked. The EPA has more information at **www .epa.gov/vaporintrusion**.

4. Use a water filter.

While we wait for cleanup, we have to protect ourselves. Water filters can help. Carbon filters attached to faucets or in water pitchers are relatively inexpensive but must be replaced regularly. They also do not extract every chemical from the water.[44] More expensive options, such as reverse osmosis systems, may remove more contaminants but require greater use of water.[45]

Unfortunately, while these filters may help filter drinking water, they do not protect us from other kinds of exposure, such as bathing. Compounds like TCE readily evaporate, and individuals with contaminated water may be exposed to them while showering. In these cases, a whole-house water-filtration system is needed to clean all the water at the point where it enters the house.[46]

To learn more, two useful resources are the EPA's Safe Drinking Water Hotline at 800-426-4791 and the EPA's webpage on groundwater and drinking water: **www.epa.gov/ground-water-and-drinking -water/safe-drinking-water-information**. In addition, the Environmental Working Group has a database of tap water contamination by zip code available at **www.ewg.org/tapwater**.

5. Advocate for the National Neurological Conditions Surveillance System.

The National Neurological Conditions Surveillance System can help identify potential causes of Parkinson's and other neurological diseases, like multiple sclerosis. However, it does not have sufficient money to do so.[47] Parkinson's can take decades to develop in response to environmental risks. Therefore, long-term funding is needed. Many of the major Parkinson's organizations have combined their lobbying efforts to support this act. You can learn more about this and other policy priorities at **https://advocate.michaeljfox.org**.

6. Protect workers.

If you are one of the 1 million farmers, ranchers, and agricultural workers in the United States, you can protect yourself by avoiding or minimizing your contact with pesticides.[48] Farmers can be exposed to pesticides during the preparation, application, and cleanup of spray solutions.[49] Agricultural workers can breathe pesticides in when they are sprayed or when they drift through the air from neighboring fields. They can also absorb the chemicals through their skin while handling crops or touching the soil.[50] Skin absorption and inhalation are the most common, but not the only, routes of exposure.[51] Paraquat, for example, "is highly toxic to animals by all routes of exposure," including via the eyes, the mouth, and even the fingernails.[52]

People working on farms need appropriate protective equipment, which can reduce the risk of Parkinson's.[53] Among these are gloves, boots, hats, long-sleeve shirts, masks with air filters, and chemical-resistant overalls. However, such protection is not always complete.[54]

Certain groups are especially vulnerable to the toxic effects of pesticides, including children, migrant workers, and residents of low-income nations. These groups may not know the importance of or have access to the proper equipment.

Industrial workers should also protect themselves from TCE and other solvents that are linked to Parkinson's.[55] The most common current uses of TCE are grease removal, dry cleaning, and the production of refrigerants. TCE is also found in consumer products including typewriter correction fluid, paint removers, adhesives, spot removers, and rug-cleaning fluids.[56] According to the Agency for Toxic Substances and Disease Registry, dry cleaners, mechanics, printers, shoemakers, textile and fabric cleaners, and varnish workers, among others, have an increased risk of exposure.[57] Industrial workers need the corresponding protection, including goggles, gloves, and respirators, until we can eliminate the dangerous product's use.[58]

Welders may also be at higher risk for developing Parkinson's.[59] These workers, who number 400,000 in the United States, should prevent or limit their exposure to fumes that result from melting the metal manganese, which is toxic to dopamine-producing nerve cells.[60] Lower welding voltages, ventilation equipment, and masks can help protect welders from inhaling the dangerous gases.

7. Eat like the Greeks.

Recent studies have suggested that a Mediterranean diet, in addition to lowering our risk of heart disease, may also be protective against Parkinson's—by as much as 20%.[61] This way of eating avoids processed foods and includes the following:

- Fruits and vegetables
- Beans and nuts
- Whole grains
- Fish
- Olive oil
- Small amounts of meat and dairy

People living in Mediterranean countries also like their coffee, and caffeine may also be protective against Parkinson's. Studies suggest that between one and four cups of caffeinated coffee may decrease your risk.[62]

8. Minimize your consumption of pesticides.

We do not have great data on the risk of eating produce coated with pesticide residues. But Dr. Honglei Chen, a professor of epidemiology and biostatistics at Michigan State University, says, "I wouldn't be surprised if regular consumptions of foods with pesticide contaminants or residues increase the risk of Parkinson's disease, particularly in countries where the use of pesticides is less regulated."

It may be prudent to take preventive measures such as washing fruits and vegetables thoroughly with a little soap and water. We might also opt for organic foods, if possible, given that they have much lower levels of pesticide residues than conventional choices.[63] We also need to work to make these organic foods more affordable and accessible to as many people as possible. For more information about which kinds of produce are sprayed the most, visit the website for the nonprofit Environmental Working Group (**www.ewg.org**), which keeps updated lists.

9. Sweat.

On average, we do not get nearly enough exercise to realize its well-documented benefits for many diseases, including Parkinson's.[64] Vigorous exercise for about four hours every week can potentially reduce the risk of Parkinson's disease by 20% or more. For people who have already developed Parkinson's, numerous studies have shown that regular exercise can alleviate symptoms.[65]

It does not matter what you do—tap dance, fly on a trapeze, weed your garden at top speed, do yoga on a mountain, or simply go out and take a brisk walk—as long as you do it. Cycling or running can be surprisingly good options even for those who find it difficult to walk normally.[66] Consistency is the key; exercise every day so there is never room to postpone it until tomorrow. Breaking a sweat is usually a good indicator that your heart rate is elevated.

Schools, workplaces, community centers, houses of worship, senior centers, and governments should all promote and enable exercise. We need to think of it as a required daily activity, like brushing our teeth. The risks are few, and the benefits are immense.

10. Avoid activities with a high risk of head trauma.

All of us should wear seat belts, drive cars with air bags, and wear helmets when riding a bike, skiing, or engaging in other activities with a high risk of concussion. Parents should exercise considerable caution in deciding whether to let their children play tackle football.[67] Athletes should also be aware of the risks. We may love the game in the United States, but it is worth thinking about what concussions may bring in the future. This applies to other sports too, though their risks may be less well publicized (**Figure 1**).

For those in the military, helmets, face shields, and improved vehicle armor can help and should be available to those at high risk of injury.[68] We need to detect the dangers and take soldiers out of harm's way before a blast occurs.[69] As with Parkinson's, a lot of current research is focused on improving the diagnosis and treatment of traumatic brain injury.[70] More is needed to prevent it.[71]

ADVOCATE FOR ADDITIONAL RESOURCES AND POLICY CHANGES

11. Increase NIH funding for Parkinson's disease.

The National Institutes of Health (NIH) is the largest public funder of biomedical research in the world. It spends $37 billion a year on research "to enhance health, lengthen life, and reduce illness and disability."[72] However, at the same time that the burden of Parkinson's is increasing, the NIH's support for Parkinson's research

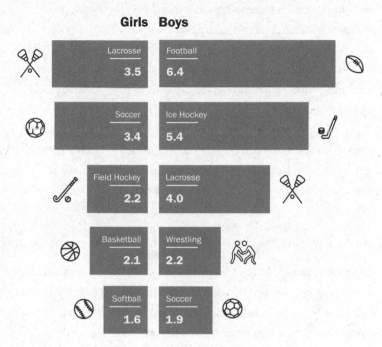

FIGURE 1. Concussion rates among high school athletes by sport per 10,000 athletic exposures, 2008–2010.[73] An athletic exposure is one player participating in one practice or game.

is lagging. More broadly, the US share of global research funding is also decreasing.[74]

More NIH funding is needed, and more NIH money needs to go toward Parkinson's research. The pie can also be expanded through strategic partnerships. Some of this work is starting. In 2018, a public-private partnership was formed to advance new treatments for Parkinson's. The NIH, multiple pharmaceutical companies, and The Michael J. Fox Foundation agreed to form the Accelerating Medicines Partnership and to invest $24 million over five years to identify promising biological markers of Parkinson's disease.[75]

Increased funding could help identify ways to reduce our risk of Parkinson's. First, we need a better understanding of the risks themselves.

How widespread are these environmental risks? What level of exposure, if any, is safe? We must also figure out how to assess this exposure in people. Can we measure these chemicals in the blood, urine, hair, nose, gut, or brain to help determine risk? Another key step will be determining how environmental and genetic risk factors interact. Are those who carry certain genetic mutations at greater risk of developing Parkinson's when exposed to certain pesticides, chemicals, or head trauma? Should these individuals be even more vigilant? Two centuries after Dr. Parkinson described the condition, we still have many unanswered questions that could help prevent this debilitating disease.

12. Listen up, Big Pharma—this is a great business opportunity.

The exploding numbers of people with Parkinson's represent an enormous unmet need—and opportunity. Alzheimer's and Parkinson's together affect at least 6 million Americans, a number that is projected to double in the next generation.[76] There is money to be made from treatments that can help them. The best treatment for Parkinson's is over fifty years old, only partially alleviates its enormous disability, and—this is key—does not address the underlying causes of the disease. Pharmaceutical companies should therefore be eager to develop new medications.

In 2018, the world's third-largest pharmaceutical company, Pfizer, announced that it was ending drug development for both conditions.[77] With this announcement, Pfizer eliminated three hundred positions from its neurodiscovery unit. Fortunately, a new spin-off company called Cerevel Therapeutics was formed to help advance treatments for Parkinson's and Alzheimer's, which also lacks effective treatments, and more companies are starting to stream back toward neurodegenerative diseases.[78] The field needs more investment from the large pharmaceutical companies, which, along with medical device companies, account for 60% of biomedical research funding in this country.[79]

13. Donate to organizations that invest in important research and care.

Tens of millions of dollars have been raised by organizations in the United States, such as The Michael J. Fox Foundation and the Parkinson's Foundation, and outside its borders—Parkinson's UK, the Cure Parkinson's Trust, and the European Parkinson's Disease Association. The funding has enabled research that has led to a better understanding of the disease, the discovery of genes, identification of drug targets, testing of new treatments, development of devices, and regulatory approval of new therapies. Money is a limiting factor for how far we can go and how fast we will get there.

The Michael J. Fox Foundation has had a large impact. The organization has no endowment and spends almost all the money it raises annually. Since its founding in 2000, it has invested over $900 million in research programs. In 2018, it spent approximately $100 million on research—more than half of what the NIH spent on Parkinson's.

More financial support remains critical, and other foundations like the American Parkinson Disease Association, the Brian Grant Foundation, the Davis Phinney Foundation, and the Parkinson's Foundation together spend more than $30 million annually. The American Parkinson Disease Association is the largest grassroots network dedicated to fighting Parkinson's and has invested nearly $200 million in patient services, educational programs, public awareness efforts, and research since 1961.[80] The Brian Grant Foundation provides exercise and nutritional resources for people with Parkinson's.[81] The Davis Phinney Foundation organizes Victory Summits around the country and around the world that educate and inspire those with Parkinson's disease to live well today. The Parkinson's Foundation and its predecessors have raised over $300 million for research and care, and the foundation now operates over forty centers of excellence worldwide that provide expert care to people with Parkinson's.[82] Many more Parkinson's disease organizations in the United States and around the world are contributing money to improve care and advance research.

A "march of dimes" funded polio research. Pink ribbons have raised millions for breast cancer. An Ice Bucket Challenge did the same for amyotrophic lateral sclerosis.[83] We need similar efforts that will translate to tangible progress for Parkinson's.

14. Be loud.

No voice is more powerful or harder to ignore than that of those affected by the condition. The Parkinson's community has been relatively quiet and patient. It has yet to find a way to demand accountability from doctors for improved care, researchers for scientific progress, pharmaceutical companies for drug development, and legislators for public health protection in a way that cannot be ignored. The HIV community was and is loud and persistent. At times, it was unsettling. Its activists even occupied the FDA's building. However, the spirit of the HIV community and its passionate drive for change drove progress.

Maybe the time has come for a Parkinson's version of ACT UP. Defined as "a diverse, non-partisan group of individuals united in anger and committed to direct action to end the AIDS crisis," the group's motto is "We advise. We demonstrate. We are not silent."[84]

15. Organize.

If all the people with Parkinson's disease flooded the National Mall in Washington, DC, they would exceed the number of people at most Presidential inaugurations. If we counted up everyone who either has the disease or loves someone who has it, our numbers would be an undeniable force. The power is there if we use it, but we must organize ourselves to be effective. All the efforts listed here and many others will require unprecedented activism.

For every one person with Parkinson's, one hundred more do not have the disease. We need to engage and organize more than just patients and care partners. We need their children and siblings, friends and neighbors, clinicians and aides to participate. All their voices need to be heard.

Many valuable Parkinson's disease organizations exist, but their impact and reach can all be increased. The March of Dimes, ACT UP, and Susan G. Komen Race for the Cure have their roots in broad, engaged, and sustained activism. That activism includes engaging those without the disease and, in particular, younger generations. These individuals have the most to gain from a cleaner environment and from ending Parkinson's.

CARE FOR ALL THOSE WHO ARE AFFECTED
16. Train more specialists and educate clinicians.

We have enough neurologists in the United States to care for people with Parkinson's. But we have few Parkinson's disease specialists. Care from this latter group, whose members devote additional training to those with the condition, can address the complex needs of many with the disease, help improve care, and advance new therapies. As more people develop Parkinson's, more such doctors will be needed. The Michael J. Fox Foundation and the Edmond J. Safra Foundation have partnered to increase the number of Parkinson's specialists primarily in the United States and Europe. The Parkinson's Foundation and its network of expert care centers has joined with the American Academy of Neurology to train more Parkinson's disease neurologists too. Despite these efforts, fewer than fifty Parkinson's specialists, on average, are trained each year in the United States. Outside the United States, the needs are far greater, particularly in middle- and low-income nations.

And we need specialists in many other professional disciplines as well. People with Parkinson's benefit from care provided by a team of specifically trained experts, including physical, occupational, and speech therapists. Dedicated Parkinson's nurses are also critically important. Specific training for these other disciplines is offered by various organizations around the world, including the International Parkinson and Movement Disorder Society, Parkinson's UK, the Parkinson's Foundation, and the Dutch ParkinsonNet. But much work remains to ensure that patients around the world have access to such experts.

We also need to train clinicians of all varieties to become more familiar with Parkinson's so people can be diagnosed more quickly—and receive treatment sooner. Raising public awareness also helps both patients and clinicians recognize symptoms earlier. And let's face it, doctors miss things all the time, even specialists like us. So people need to make sure they are persisting, pushing for answers, and receiving multiple opinions if they know that something is wrong.

17. Expand access to care.

Today, we are failing to take proper care of people with Parkinson's. The way we provide care now restricts access to a fortunate few and is overly burdensome to patients and their caregivers. We can and must do better.

ParkinsonNet in the Netherlands, the network of specially trained experts in Parkinson's, offers an excellent example of how to expand access to care.[85] That model is now spreading to other countries in Europe and even to the United States but is still not available to all.

Other ways to take better care of patients are also emerging. For example, the "Service and Science" hub at the University of Florida, headed up by Dr. Michael Okun, one of our authors, provides one-stop shopping for patients—medical attention, support, and the ability to participate in clinical studies all in one location.[86] This model could be replicated at many locations both inside and outside the United States, but large-scale funding remains sparse.

Another model is Project ECHO, which uses remote specialists to train and support local clinicians. Even though the model is low-cost and builds local capacity, it is not currently covered by most insurers.[87]

18. Lobby to change insurance and Medicare.

Decisions by health insurers and Medicare about which services they are willing to cover determine what kind of care we receive. For people with Parkinson's, Medicare will pay for institutional care—hospital stays, nursing homes—but very little home care. This is despite the fact that most people would rather stay in their own homes and that home care is far cheaper.

One way to ensure that Medicare's policies are consistent with the interests of its beneficiaries is to expand the physician-only advisory committee to include patients. Giving them a voice will help ensure closer alignment of Medicare's financial incentives with the interests of those who stand to gain and lose the most. People from the Parkinson's community need a seat at the table.

19. Enable individuals with Parkinson's to live at home.

Several services can help people live at home and postpone or minimize the need for a nursing home. The first is house calls, which can be delivered in person or via videoconferencing. Another is in-home exercise programs that are taught or led by Parkinson's trained therapists, which can improve function for people with the disease. For example, people can now ride a stationary bicycle at home with remote supervision by an exercise coach.[88] A third is a novel program that dispatches a nurse, an occupational therapist, and a handyman to the homes of older adults. There they provide assistive devices and repair and modify homes. This simple intervention reduces depression and increases the independence of older adults.[89] The obvious hurdle to widespread adoption of these and similar programs is the lack of Medicare and Medicaid reimbursement. If we can make policy changes, the savings in cost and heartache could be billions of dollars and millions of lives.

The demand for programs like these will only grow as more people live with advanced Parkinson's disease. Currently, 2 million Medicare beneficiaries are homebound, leaving their home less than once a week, often only to see a doctor. Of these 2 million, tens of thousands likely have Parkinson's.[90] For people who are homebound, Medicare may provide home health coverage. However, the rules are complex, and the care does not always cover the needed services. However, for the vast majority of individuals with Parkinson's who are not homebound, Medicare's coverage is limited.[91] Spouses and other loved ones end up having to perform this monumental and stressful job.

20. Use technology to increase access to care.

While most individuals lack ready access to a neurologist globally, 90% of them have a mobile phone, and 70% have a smartphone.[92] Because mobile phone ownership is high even in low-income nations, these devices can help extend care. Coupling this technology with remote support through telemedicine links can extend appropriate care to almost anyone anywhere. This is already happening in India, where physicians are seeing patients with Parkinson's disease through a smartphone.

Much more is possible. In the near future, smartphones with their embedded sensors will facilitate remote diagnoses and monitoring.[93] The same devices that have transformed communication can now transform care.

TREAT PARKINSON'S DISEASE WITH EFFECTIVE THERAPIES
21. Make dopamine-replacement therapy more widely available.

Though levodopa is inexpensive in high-income nations, infrastructure and bureaucratic issues keep this generic drug from reaching people in many parts of the world. The World Health Organization includes levodopa on its list of essential medicines, yet access is largely limited to high-income nations.[94] There is no excuse for this. President George W. Bush championed providing HIV drugs to people in Africa despite the fact that these drugs are far more expensive than simple dopamine replacements. We need similar leadership for levodopa and Parkinson's disease.

22. Participate in clinical trials.

Promising therapies are in the pipeline for Parkinson's. However, all these therapies require extensive evaluation in clinical trials, and almost 80% of these studies fail to meet their targets for recruitment.[95] We cannot get closer to having better therapies or finding a cure without broad and enthusiastic participation from the Parkinson's community.

We should seek to identify potentially eligible participants before a clinical trial even begins. The community can do this by signing up for observational research studies that track individuals with a genetic mutation, for example, before gene-directed trials start. Other resources, such as Fox Trial Finder, can readily match individuals with Parkinson's to clinical studies in which they can participate.[96] The faster trials complete enrollment, the quicker they are finished, the sooner the results are known, and the closer we get to new treatments.

23. Take research studies to participants.

Clinical trials depend on people who are willing to donate their time and energy, to share their body fluids, and to expose themselves to known and unknown health risks. Yet we burden these volunteers with traveling to research sites in order to participate. We can invert this model and take the research to participants. Doing so will lessen the strain on the people already grappling with a disease, and science would benefit.

Currently, only 10% of people with Parkinson's participate in research. Extending the reach to underrepresented populations will enhance the generalizability of findings and ultimately benefit more people.[97] And the larger the numbers of participants, the stronger the evidence.

Early studies indicate that there is interest in this novel approach to doing research. A study that allowed people to sign up on their smartphones enrolled 15,000 participants, including people from every state.[98] That is a huge number, and it was because the participants did not have to travel anywhere.

More recently, The Michael J. Fox Foundation created Fox Insight, an online research platform that enables individuals to participate in research without ever having to leave their home.[99] To date, over 40,000 individuals have enrolled.

24. Study people with early disease symptoms and those who are at greatest risk.

By the time Parkinson's is diagnosed, much of the damage is largely done—two-thirds of the dopamine-producing nerve cells in the brain are already lost. We need to study people earlier in the course of Parkinson's. This is critical to learning how the disease progresses in its earliest stages and what might stop it.

We should be reaching out to people who are experiencing very early symptoms, such as loss of smell, but who have not yet been diagnosed. The International Parkinson and Movement Disorder Society has developed a tool that can help identify such individuals.[100] We can also look at populations who are at high risk because of genetic factors or environmental exposures. Some of this work is already underway, and our understanding will greatly benefit from enrolling many more people in these early-stage efforts.

25. Provide reasonable pricing of Parkinson's drugs.

In Europe, government agencies pay for therapies based on their value.[101] While the United States lacks such regulatory bodies, we should also be tying price to effectiveness. For example, new models in cancer are beginning to emerge that require insurers to pay for expensive drugs only if patients respond to treatment.[102]

While high-income nations may be able to afford expensive therapies, middle- and low-income nations cannot. For some of these countries, the price may have to be less than the cost to manufacture drugs. Such an approach for people with HIV/AIDS has saved millions of lives and, overall, has made billions for pharmaceutical companies. A similar model could be developed for Parkinson's.[103]

These actions are achievable. Most have little, if any, cost. Most have little, if any, downside. They all have enormous benefits. If we pursue this prescription, fewer of us will live our last decades with a debilitating disease or spend years caring for someone with Parkinson's.

ठ ठ ठ

To a great extent, Parkinson's may be man-made. Pesticides, chemicals, air pollution, head trauma, and lack of exercise are all fueling the rise of the disease and putting us all at risk. Just as humans contributed to the rise of Parkinson's in the nineteenth and twentieth centuries, we can now work to eradicate the disease.

Previous generations have eliminated infectious diseases like smallpox and polio. As a result, we no longer fear or even think about the scars, paralysis, and death that these viruses wrought.

And we have kept fighting ever since. The greatest medical advance of the last generation was the transformation of HIV/AIDS from an unknown, fatal condition into a preventable and treatable one. That shift was not easy; it never is. But we now know it's possible, and we know what it takes. People had to throw out the status quo, end their silence, and act.

It's now on us to do the same for Parkinson's disease. If we succeed, future generations will not have to confront the hardship and disability that Parkinson's brings. If we fail, a needless pandemic will be our legacy.

What will be our story?

POSTSCRIPT

Unless commitment is made, there are only promises and hopes…but no plans.

—Peter Drucker, management consultant,
educator, and writer

ENDING PARKINSON'S DISEASE WAS RELEASED ON MARCH 17, 2020, in the midst of a global pandemic of a novel virus. The large scope of this disease has overshadowed the insidious spread of another entity that affects millions and that will persist long after COVID-19 disappears—the Parkinson's pandemic. However, the global effort to address COVID-19 has shown how effective the PACT—prevention, advocacy, care, and treatment—could be to end the world's fastest-growing brain disease.

Since the book's publication, some progress has been made against Parkinson's, including increased research funding from the National Institutes of Health (NIH), the formation of an international grass-roots movement, and new treatments. But despite these early efforts, we need to be far more ambitious.

The first step to addressing any pandemic is to contain it. To slow the spread of COVID-19, we initially had to recognize its threat so we could implement measures such as physical distancing, masks, and widespread testing. Countries that acted early, aggressively, and consistently, like New Zealand and Taiwan, prevented many of their citizens from ever becoming sick. Countries that ignored or down-played the threat saw large numbers of infections and deaths.[1]

Unfortunately, we have failed to contain the Parkinson's pandemic; instead, we are fueling it. Paraquat continues to be sprayed unabated on farms in the U.S., and its use has *tripled* over the past twenty-five years.[2] This is how Parkinson's disease grows. Despite the chemical's dangers, which have led thirty-two countries to ban it, the U.S. Environmental Protection Agency (EPA) reapproved the use of paraquat at the end of 2020.[3] In addition, hundreds of sites remain contaminated with the industrial solvent trichloroethylene (TCE). And more clusters of Parkinson's near these and other polluted areas are emerging.

Determining where these danger zones can help protect people from them. To the dismay of many, the National Library of Medicine recently took down TOXMAP, one of the key resources that identified areas contaminated with TCE and other pollutants.[4] This map allowed scientists, advocates, and citizens to pinpoint and evaluate potential drivers of Parkinson's disease, cancer, and other conditions. Yet, the U.S. government dismantled a valuable means of tracking not only TCE but also air and water pollution that are contributing to the rise of Parkinson's and Alzheimer's disease as well.[5] To slow the Parkinson's pandemic, we need more, not fewer, tools to assess the effects of pesticides, solvents, heavy metals, and air pollution. Once identified, we need to eradicate these toxins from our environment.

On the advocacy front, we have made some headway. While the U.S. is permitting continued spraying of dangerous pesticides, Europe is close to ending its use of maneb, another pesticide linked to Parkinson's.[6] Also in 2020, U.S. Senator Tom Udall and Congressman Joe Neguse introduced landmark legislation to prevent the use of toxic pesticides that harm children, farmworkers, and consumers. The Protect America's Children from Toxic Pesticides Act represents the first comprehensive update since 1996 to the law governing the use of pesticides. Among other things, the legislation would ban two pesticides linked to Parkinson's, paraquat and chlorpyrifos, and correct the EPA's recent mistake in allowing paraquat's continued use.[7] In addition to its effect on Parkinson's, chlorpyrifos impairs brain development and has robbed children of seventeen million IQ points. As we write this piece, the prospects for the bill's passage are limited. Its introduction, however, will pave the way for future action aimed at lowering our collective risk of developing Parkinson's disease.

In addition to removing threats that we know raise the risk of Parkinson's, we need more data on potential emerging factors. A new and disconcerting one may be COVID-19 itself. This virus is damaging not just our lungs, but also possibly our brains.[8] As of this writing, a small handful of cases of Parkinson's disease have been associated with COVID-19 infection.[9] This link requires immediate study and careful tracking.

COVID-19 has also shown how financial resources can make a difference in addressing a pandemic. Up to $2.6 trillion including $250 billion for the U.S. Department of Health and Human Services have been devoted to addressing the COVID-19 pandemic.[10] This funding is resulting in improved diagnostic testing, robust vaccine development, and potential new therapies. We need to bolster our investment in Parkinson's, too. In 2019, NIH funding for Parkinson's research did rise to $224 million, which is up 40% from $161 million in 2016.[11] However, that funding will have to increase tenfold or more to move the needle against this debilitating disease.

We are not alone in advocating for more Parkinson's-protective policies. We are humbled to report that after reading our book, a new advocacy group called the PD Avengers was inspired to take action. The organization is led by Larry Gifford, host of the podcast *When Life Gives You Parkinson's*. An Ohio native now living in Vancouver, Canada, Gifford was diagnosed with Parkinson's in 2017 at the age of forty-five. The Avengers' global campaign aims to have one million individuals sign the Parkinson's PACT to from a coalition that advocates for equitable access to therapies and looks to eliminate Parkinson's through research and prevention. You can join the first signatories at www.pdavengers.org.

In the silver lining department, COVID-19 has increased access to care for some individuals with Parkinson's. On the very day that our book was published, the Centers for Medicare & Medicaid Services announced the vast expansion of telemedicine coverage to enable older adults to receive care without risk of infection. Medicare beneficiaries, including many with Parkinson's, immediately embraced this option, and in less than a month, telemedicine appointments increased more than one-hundred-fold.[12]

This long-desired change, which allows many to receive care at home, may not last. Because it is tied to the current Public Health Emergency, the expanded coverage is only temporary. To ensure that individuals with Parkinson's disease, COVID-19, and other conditions continue to receive such care, long-term policy changes will be required.

Telemedicine is not a U.S. phenomenon. Patients in Italy, Germany, and the Netherlands, for example, have all continued to receive care without the threat of COVID-19 thanks to phone calls with nurses and video calls with neurologists and geriatricians.[13] In China, individuals with deep brain stimulators have benefited from new wireless technology that allows for remote adjustments.

As much as telemedicine can facilitate access to care, it is not a panacea. Social and economic factors continue to limit who can take advantage of it. Individuals who are older, less educated, or live in rural areas often cannot and do not benefit from remote visits.[14] Because people of color often lack access to health care resources, they are less likely to see a Parkinson's specialist, and consequently, many people in this group go undiagnosed.[15]

Internet connectivity is another limiting factor. COVID-19 demonstrated how the availability of WiFi can determine, for example, which children receive an education.[16] Treating internet access as a universal public right is part of our vision of supporting people living with Parkinson's. Anyone, anywhere, with Parkinson's should be able to receive care.

Finally, better treatments are also needed for the millions who are living with Parkinson's today.

COVID-19 has demonstrated how a sense of urgency and massive funding rapidly led to novel therapies. We must develop a diversified portfolio of Parkinson's treatments, and our remedies must be tailored to

There are an estimated

1,171,301

people diagnosed with **Parkinson's disease** in the U.S. right now.

Every 7 minutes a new person is
diagnosed with **Parkinson's disease**
in the U.S.

FIGURE 1. Number of Americans diagnosed with Parkinson's disease as of October 27, 2020. A current estimate is available at www.pdclock.org. The worldwide prevalence is likely six to ten times greater.

the individual. Since our book was written, a handful of medications—istradefylline, opicapone, and a new formulation of apomorphine—have been approved. However, these are only incremental advances. We must build a better pipeline for drugs that eventually will alleviate symptoms, slow disease progression, and ideally in some cases, even prevent Parkinson's. To that end, we need the Parkinson's community engaged in clinical trials, education, and public policy.

The Parkinson's pandemic has exacted an enormous toll on tens of millions of individuals who bear the burden of the disease. Hundreds more are diagnosed every day. Two centuries after its seminal description, the time has come to alter its course. To prevent and end this disease, we must be aggressive, charismatic, bold, and willing to change the status quo. We hope that *Ending Parkinson's Disease* provides a prescription for action and success on that journey.

A PRESCRIPTION
for ACTION

J.W.S. van der Wereld, a well-known Dutch horticulturalist who had
Parkinson's disease, named a red tulip with fringes of white as the
"Dr. James Parkinson" tulip.

The tulip is now an international symbol that represents the community's hope
and optimism for a world without Parkinson's disease.

PARKINSON'S DISEASE IS THE FASTEST

GROWING NEUROLOGICAL DISORDER

IN THE WORLD.

WE CAN END IT.

THE FOLLOWING MEASURES FORM
A **PACT** TO HELP END PARKINSON'S.

PREVENT

ADVOCATE

CARE

TREAT

PREVENTING PARKINSON'S

Ban paraquat and trichloroethylene

Contact the EPA Administrator, currently Andrew Wheeler, and ask him to ban the pesticide paraquat and the chemical trichloroethylene.

E-MAIL THE EPA ADMINISTRATOR
WHEELER.ANDREW@EPA.GOV

CALL HIM
(202) 564 – 4700

Drink clean water

Test your water, especially if you have a well.
Use water filters to reduce exposure to chemicals.

EPA SAFE DRINKING WATER HOTLINE
(800) 426 – 4791

EPA'S WEBSITE ON PRIVATE WELLS
EPA.GOV/PRIVATEWELLS/

Breathe fresh air

See if you live near a trichloroethylene-contaminated site and test your air.

LEARN ABOUT SUPERFUND SITES NEAR YOU
EPA.GOV/SUPERFUND/SEARCH-SUPERFUND-SITES-WHERE-YOU-LIVE

CONTACT THE EPA ABOUT INDOOR AIR QUALITY
EPA.GOV/VAPORINTRUSION

Eat healthy

Avoid contaminants in food. Thoroughly wash foods that may be contaminated with pesticides linked to Parkinson's, and consider organic sources.

LEARN MORE ABOUT PESTICIDES IN PRODUCE
EWG.ORG/FOODNEWS

Reduce your risk at work

Minimize exposure at work. If you work with pesticides, trichloroethylene, or other harmful chemicals, wear masks, gloves, and protective clothing.

NATIONAL PESTICIDE CENTER WEBSITE   **NATIONAL PESTICIDE CENTER HOTLINE**
NPIC.ORST.EDU/REG/WPS.HTML   (800) 858 – 7378

Exercise, eat well, and enjoy your coffee

Exercise vigorously, eat a Mediterranean diet, and consume modest amounts of caffeine.

Avoid activities with a high risk of concussion

Wear a protective helmet when participating in such activities.

PACT

ADVOCATING FOR PARKINSON'S

Call your Representative and push for more funding

Ask your Representative and Senators to join the bipartisan Congressional Caucus on Parkinson's disease. Tell them to increase NIH funding for Parkinson's research, which trails the increasing burden.

VISIT THE PARKINSON'S CAUCUS SITE
PARKINSONSCAUCUS.ORG

Share your story

Find and attend a Parkinson's support group, either in-person or online.

APDA CHAPTERS AND INFORMATION & REFERRAL CENTERS
APDAPARKINSON.ORG/COMMUNITY

DAVIS PHINNEY FOUNDATION MOMENTS OF VICTORY®
DAVISPHINNEYFOUNDATION.ORG/GET-CONNECTED/MOMENTS-OF-VICTORY/

FACEBOOK GROUPS
EVERYTHING ABOUT PARKINSON'S DISEASE
LIFE WITH PARKINSON'S
PARKINSON'S COMMUNITY
THE PARKINSON'S EXPERIENCE SUPPORT GROUP
PARKINSON'S PALS
STRONGHER: WOMEN FIGHTING PARKINSON'S

PARKINSON'S FOUNDATION CHAPTERS
PARKINSON.ORG/CHAPTERS

Stay informed

Sign up to receive policy alerts and updates.

WEBSITE
ADVOCATE.MICHAELJFOX.ORG

Support research efforts

Contribute time or money to a Parkinson's organization and hold researchers accountable.

PARKINSON'S UK
CHANGE ATTITUDES.
FIND A CURE.
JOIN US.

THE MICHAEL J. FOX FOUNDATION
FOR PARKINSON'S RESEARCH

AND MANY MORE
GREAT ORGANIZATIONS...

PACT

CARING FOR PEOPLE

WITH PARKINSON'S

Visit a center of excellence

Go to a Parkinson's disease center to receive expert care.

APDA ADVANCED CENTERS
APDAPARKINSON.ORG/RESEARCH/ADVANCED-CENTERS/

PARKINSON'S FOUNDATION CENTER OF EXCELLENCE NETWORK
PARKINSON.ORG/SEARCH

Receive the care you want

Contact the Secretary of Health and Human Services, currently Alex Azar, and ask him to include the voices of patients in determining how Medicare spends its money.

E-MAIL HIM
SECRETARY@HHS.GOV OR ALEX.AZAR@HHS.GOV

Expand Medicare's coverage of telemedicine

Tell your Representative or Senator to make the temporary changes expanding Medicare's coverage of telemedicine permanent. You can also contact the Administrator of the Centers for Medicare & Medicaid Services, currently Seema Verma.

E-MAIL HER
SEEMA.VERMA@CMS.HHS.GOV

SEND HER A TWEET
@SEEMACMS

Learn more about Parkinson's

Watch ParkinsonTV episodes and go through our resources section.

DAVIS PHINNEY FOUNDATION BLOG AND WEBINARS
DAVISPHINNEYFOUNDATION.ORG/BLOG/

MICHAEL J. FOX FOUNDATION WEBINARS
MICHAELJFOX.ORG/WEBINARS

PARKINSON'S FOUNDATION LIBRARY
PARKINSON.ORG/LIBRARY

PARKINSONNET
PARKINSONNET.COM

PARKINSONTV
PARKINSONTV.ORG

PARKINSON'S UK SUPPORT HUB
PARKINSONS.ORG.UK/INFORMATION-AND-SUPPORT

PACT

TREATING

PARKINSON'S

Join a clinical study

Participate in clinical trials of new therapies.

TRIAL REGISTRY
CLINICALTRIALS.GOV

FOX TRIAL FINDER
FOXTRIALFINDER.MICHAELJFOX.ORG

PARKINSON NEXT (NETHERLANDS)
PARKINSONNEXT.NL

PARKINSON'S UK TAKE PART HUB
WWW.PARKINSONS.ORG.UK/RESEARCH/TAKE-PART-RESEARCH

Consider your family history

Talk to your doctor and a genetic counselor about the pros and cons of genetic testing.

FIND A GENETIC COUNSELOR
WWW.NSGC.ORG/PAGE/FIND-A-GENETIC-COUNSELOR

PARKINSON'S FOUNDATION GENETIC STUDY
PARKINSON.ORG/PDGENERATION

Participate in research from your home

Enroll in online clinical studies.

FOX INSIGHT

FOXINSIGHT.MICHAELJFOX.ORG

ALL OF US RESEARCH PROGRAM

JOINALLOFUS.ORG

PACT

RESOURCES

For updates, additional
resources, and current advocacy
information, visit
endingPD.org

We welcome your feedback at
info@endingPD.org

Prevent

LAND

Agricultural Pesticide Use
water.usgs.gov/nawqa/pnsp/usage/maps
VIEW THE PARAQUAT MAP

Center for Public Environmental Oversight
cpeo.org

Superfund
epa.gov/superfund

TOXMAP—taken down in 2019
toxmap.nlm.nih.gov/toxmap
SEARCH FOR TRICHLOROETHYLENE

WATER

Tap Water Pollutants
ewg.org/tapwater
ENTER YOUR ZIP CODE

EXERCISE

American Parkinson Disease Association Resource Center Exercise Helpline
(888) 606 – 1688
rehab@bu.edu

BurnAlong
burnalong.com

Dance for PD
danceforparkinsons.org

Davis Phinney Foundation Parkinson's Exercises Essentials Video
davisphinneyfoundation.org/parkinsons-exercise-essentials-video

Parkinson's UK Exercise Recommendations
parkinsons.org.uk/information-and-support/exercise

PD Movement Lab
pdmovementlab.com

Rock Steady Boxing
rocksteadyboxing.org

110 Fitness
110fitness.org

Advocate

ADVOCATES

PD Avengers
pdavengers.com

LEGISLATION

Michael J. Fox Foundation Advocacy Toolkit
michaeljfox.org/policy

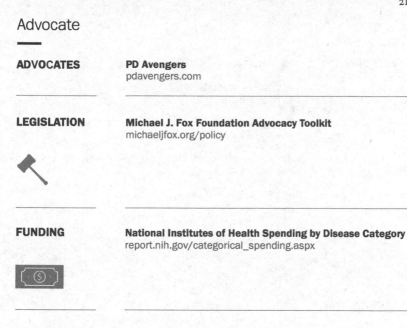

FUNDING

National Institutes of Health Spending by Disease Category
report.nih.gov/categorical_spending.aspx

ENGAGEMENT

Advocating for Parkinson's in the UK
parkinsons.org.uk/get-involved/
join-our-campaigns-network-today

Connect with a Davis Phinney Foundation Ambassador
davisphinneyfoundation.org/connect

World Parkinson Coalition
worldpdcoalition.org

World Parkinson Congress
worldpdcongress.org
wpc2022.org

Care

HELPLINES

American Parkinson Disease Association
1 (800) 223 – 2732
apda@apdaparkinson.org
apdaparkinson.org

Parkinson Canada
1 (800) 565 - 3000
info@parkinson.ca

Parkinson's Foundation
(800) 4PD – INFO
helpline@parkinson.org
parkinson.org

Parkinson's UK
0808 – 800 – 003
parkinsons.org.uk

FOUNDATIONS

American Parkinson Disease Association
apdaparkinson.org

Brian Grant Foundation
briangrant.org

The Cure Parkinson's Trust
cureparkinsons.org.uk

Davis Phinney Foundation
davisphinneyfoundation.org

Dutch Parkinson's Disease Association
parkinson-vereniging.nl

European Parkinson's Disease Association
epda.eu.com

The Michael J. Fox Foundation
michaeljfox.org

Parkinson's Australia
parkinsons.org.au

Parkinson Canada
parkinson.ca

Parkinson's Foundation
parkinson.org

ParkinsonNet
parkinsonnet.com

Parkinson's Society Nova Scotia
parkinsonsocietynovascotia.com

Parkinson's UK
parkinson.org.uk

Shake It Up Australia Foundation
shakeitup.org.au

**Additional foundations
around the world**
movementdisorders.org/MDS/
Resources/Helpful-Links.htm

Care

___

EDUCATION

American Parkinson Disease Association Fitness Training
apdaparkinson.org/pd-fitness-training/

American Parkinson Disease Association Publications
apdaparkinson.org/resources-support/download-publications

Every Victory Counts Manual
davisphinneyfoundation.org/resources/every-victory-counts-manual

Michael J. Fox Foundation Webinars
michaeljfox.org/webinars

Parkinson's Foundation Webinars
parkinson.org/expert-care/Professional-Education/Webinars

ParkinsonTV
parkinsontv.org

Parkylife
parkylife.com

Parkinson's Treatment: 10 Secrets to a Happier Life
by Michael S. Okun
Available on Amazon and Barnes & Noble

Parkinson's Foundation Library, Podcast & Webinars
parkinson.org/pd-library
parkinson.org/Living-with-Parkinsons/Resources-and-Support/Podcast
parkinson.org/expert-care/Professional-Education/Webinars

TRAINING

Parkinson's Academy
parkinsonsacademy.co

ParkinsonNet
parkinsonnet.com

SPECIALISTS

Florida Parkinson Foundation Center of Excellence
movementdisorders.ufhealth.org/for-patients/clinics/
parkinsons-disease-center-of-excellence/

Parkinson's Disease Care New York
pdcny.org

Parkinson's Foundation Centers of Excellence
parkinson.org/search

Parkinson Healthcare Finder (Netherlands)
parkinsonzorgzoeker.nl

Parkinson's Disease Research, Education and Clinical Center
at the U.S. Department of Veterans Affairs
parkinsons.va.gov

Treat

SCIENCE

10 Breakthrough Therapies in Parkinson's Disease
by Michael S. Okun
Available on Amazon and Barnes & Noble

***Brain Storms* by Jon Palfreman**
Available on Amazon and Barnes & Noble

International Parkinson and Movement Disorders Society
movementdisorders.org

Parkinson's Foundation Science News
parkinson.org/blog

Science of Parkinson's Blog
scienceofparkinsons.com

University of Rochester Neurology Study Interest Registry
neurologyregistry.org

University of Rochester PARK Study
parktest.net

STUDIES

ClinicalTrials.Gov
clinicaltrials.gov

Fox Insight
foxinsight.org

Fox Trial Finder
foxtrialfinder.michaeljfox.org

Morris K. Udall Centers of Excellence
ninds.nih.gov/current-research/focus-research/
focus-parkinsons-disease/udall-centers

mPower Smartphone App
parkinsonmpower.org

ParkinsonNEXT (Netherlands)
parkinsonnext.nl

Parkinson's UK Research Hub
parkinsons.org.uk/research/take-part-research

Disclosures

THREE OF US ARE ACADEMIC RESEARCHERS AND CLINICIANS who rely on the generosity and support of funders for our work. In addition, we serve as consultants to companies and organizations that are engaged in Parkinson's disease advocacy, care, education, or research. These relationships undoubtedly influence our perspectives. We have sought to minimize the potential for bias in our writing of *Ending Parkinson's Disease*. For transparency, we disclose our relationships since 2017 here.

Ray Dorsey has served as a consultant to Abbott, AbbVie, American Well, Biogen, Clintrex, DeciBio, Denali Therapeutics, GlaxoSmithKline, Grand Rounds, Huntington Study Group, Mednick Associates, Michael J. Fox Foundation, Neurocrine, Olson Research Group, Origent, Pear Therapeutics, Prilenia, Putnam Associates, Roche, Sanofi, Sunovion Pharmaceuticals, and Voyager Therapeutics. He has received grant or research support from AbbVie, Acadia Pharmaceuticals, AMC Health, Biogen, Biosensics, Burroughs Wellcome Fund, Duke University, Food and Drug Administration, Greater Rochester Health Foundation, Huntington Study Group, Michael J. Fox Foundation, NIH, Nuredis Pharmaceuticals,

Patient-Centered Outcomes Research Institute, Pfizer, Safra Foundation, Sage Bionetworks, and the University of California, Irvine. He receives compensation from Karger Publications for serving as the editor of the journal *Digital Biomarkers*. He has ownership interests in Grand Rounds (a second-opinion service).

Michael Okun serves as the medical director for the Parkinson's Foundation, is an associate editor for *JAMA Neurology* and the *New England Journal of Medicine Journal Watch Neurology*. He has received research grants from the Bachmann-Strauss Foundation, Michael J. Fox Foundation, NIH, Parkinson Alliance, Parkinson's Foundation, Smallwood Foundation, Tourette Association of America, and UF Foundation. His deep brain stimulation research is supported by two grants from the NIH (R01NR014852 and R01NS096008), and he is on mentorship committees for several NIH career-development awards. He also has received royalties for publications for books on Parkinson's disease and other movement disorders from Amazon, Books4Patients, Cambridge, Demos, Manson, and Smashwords. He has participated in continuing medical education and other educational activities on movement disorders sponsored by the American Academy of Neurology, Henry Stewart, the International Parkinson Disease and Movement Disorder Society, MedEdicus, MedNet, PeerView, Prime, QuantiaMD, Vanderbilt University, and WebMD/Medscape. The University of Florida receives grants from Abbott, AbbVie, Allergan, Boston Scientific, and Medtronic for work that Dr. Okun helps lead. He has also participated in, but not received honoraria for his role as an investigator for, several NIH-, foundation-, and industry-sponsored clinical trials.

Bas Bloem has received honoraria for serving on the scientific advisory board for AbbVie, Biogen, UCB, Walk with Path, and Zambon, has received fees for speaking at conferences from AbbVie, Bial, GE Healthcare, Roche, and Zambon, and has received research support from the Netherlands Organization for Scientific Research, Michael J. Fox Foundation, UCB, AbbVie, Stichting ParkinsonFonds, Hersenstichting Nederland, Stichting Parkinson Nederland, Parkinson's Foundation, Verily Life Sciences, Horizon 2020, Topsector Life Sciences and Health, and Parkinson Vereniging.

Todd Sherer has nothing to disclose.

Acknowledgments

ENDING PARKINSON'S DISEASE WAS A TRUE TEAM EFFORT. IT took four authors supported by a cadre of contributors, experts, readers, and institutions to make the book a reality. We thank each of them. For those that we miss here, we apologize and look forward to thanking you in person and on our website (www.endingPD.org).

Numerous individuals contributed their stories to this book and its development. They include Jay Alberts, Kevin Biglan, Trudy Bloem, Clifford Boothe, Eric Caine, Jimmy Choi, Lucien Côté, Scott DeHollander, Bob Dein, John Dorsey, Frida Falcon, Marcus Falcon, Lori Lou Freshwater, Danny Fromm, Don Gash, Jason Harvey, Jane Horton, Lyndsey Isaacs, Myra Kooy, Hyam Kramer, Alan Leffler, Max Little, Terri McGrath, Patti Meese, Eli Pollard, Judy Rosner, Jonathan Silverstein, Richard Stewart, Chuck Vandenberg, Bob van Gelder, Guy Wilcox, and many others who are not named. They all made the book come alive.

To ensure the book was accurate, we asked many friends, colleagues, and experts from a wide range of disciplines to review our work or to make suggestions. Some of them are the authors of the studies featured in the book. We thank Roy Alcalay, Alberto Ascherio, James Beck, Karen Berger, Gretchen Birbeck, Cynthia Boyd, Heiko Braak,

Honglei Chen, Nabila Dahodwala, Polly Dawkins, Alexis Elbaz, Victor Fuchs, Rebecca Gilbert, Tim Greenamyre, Christine Hay, Karl Kieburtz, Allison Kurian, Samantha King, William Langston, John Lehr, Hamilton "Chip" Moses, Marten Munneke, Jon Palfreman, Leilani Pearl, Bart Post, Briana Rae, Deborah Slechta, Katrina Smith Korfmacher, Sara Riggare, Ira Shoulson, Lenny Siegel, Caroline Tanner, Allison Willis, and others. Not everyone agreed with every conclusion we reached, but all provided valuable guidance and added rigor to the book's development.

We also thank the US Environmental Protection Agency for providing data on Superfund sites contaminated with trichloroethylene.

Ending Parkinson's Disease also benefitted immensely from the critical review of countless readers from within and outside the Parkinson's community. They had the patience to plough through far more technical earlier drafts so you would not have to. We are indebted to Jeffrey Alexis, Paul Cannon, Robert Cohen, Steve DeMello, Chris Hartman, Chad Heatwole, Jeff Hoffman, Dan Kinel, John Markman, Victor Poleshuck, Rich Simone, Pamela Quinn, and Mark Zupan. They all gave their time and insights generously.

The talented people we work with at our great institutions helped shape, edit, research, reference, depict, and develop policy recommendations that contributed immensely to the book. At the University of Rochester, we thank Olivia Brumfield, Gerardo Torres Davila, Alistair Glidden, Reenie Marcello, Taylor Myers, Karen Rabinowitz, Kelsey Spear, Anna Stevenson, Emma Waddell, and Ellen Wagner. We also thank Bryan Ingram and Monica Piraino from Brand & Butter. At The Michael J. Fox Foundation, we thank Krishna Knabe, Maggie McGuire Kuhl, and David Lubitz. At the University of Florida, we thank Polly Glattli, Melissa Himes, Leilani Johnson, Shuri Pass, and A. J. Yarbrough. And at Radboud University Medical Center in Nijmegen, the Netherlands, we thank Hanneke Kalf, Maarten Nijkrake, and Ingrid Sturkenboom. We are especially grateful to Gerardo, Emma, and Ellen for all their work developing the book's visuals, managing countless drafts, and incorporating hundreds of references.

Ray Dorsey thanks the University of Rochester for a sabbatical to write this book and the late David M. Levy and his family, the King family, and many others for their generosity. He thanks his colleagues at the Center for Health + Technology for their patience while he wrote and his chair, Dr. Robert Holloway, for fifteen years of unfailing support. He is grateful to the National Institute of Neurological Disorders for funding a Udall Center (grant P50 NS108676) to develop novel approaches to accelerate Parkinson's research. The Udall Centers fund Parkinson's research throughout the United States. They are named in memory of Congressman Morris K. Udall, who served in Congress for thirty years, the last twelve after he was diagnosed with Parkinson's disease.

Todd Sherer thanks all the supporters of The Michael J. Fox Foundation for their commitment to the mission of seeking a cure for Parkinson's disease. He is grateful for the advice and dedication of The Michael J. Fox Foundation's Board of Directors, Patient Council, and Scientific Advisors. He would especially like to thank the foundation's co-founders, Michael J. Fox and Deborah Brooks, for all that they do on behalf of the Parkinson's community and for leading the charge to develop new treatments for Parkinson's. At the foundation, he acknowledges the long-standing contributions of his colleagues Sohini Chowdhury, Brian Fiske, Mark Frasier, and Holly Teichholz, who have all spent countless hours over the past decade and a half to improve the lives of those with Parkinson's disease.

Michael Okun thanks Kelly Foote and the neurosurgery team, Nick McFarland and the movement disorders neurology team, Lisa Warren and the rehabilitation team, Herb Ward, Dawn Bowers, and the psychiatry and neuropsychology team, as well as Karen Hegland and the speech and swallow team. Without the partnership and support of this interdisciplinary group, working on the book would not have been possible.

Bas Bloem thanks everyone on his amazing team at the Center of Expertise for Parkinson & Movement Disorders at the Radboud University Medical Centre in Nijmegen; a special word of thanks goes to Marten Munneke, who is the co-architect of all the innovations dedicated to Parkinson's coming from the Nijmegen group. A big thank you to the neurologists (Rianne Esselink, Rick Helmich, Bart Post, Peter Praamstra, Suzanne ten Holter, Hanneke Thoonsen, Monique Timmer, and Bart van de Warrenburg), the Parkinson's nurse specialists (Jacqueline Deenen, Anke Elbers, Hendriette Faassen, Chyntia Geutjes, Martha Huvenaars, Myriam Koster, and Berna Rood), and all the other members of the multidisciplinary clinical team who work relentlessly every day to provide the best possible quality of life to people with Parkinson's and their families. A final word of thanks goes to Mark Tiemessen, who directs the Dutch ParkinsonNet and to Lonneke Rompen and Sanne Bouwman, who leads the ParkinsonNet international team.

The book benefited greatly from Aspen Words, whose team (Adrienne Brodeur, Marie Chan), teachers (Helen Schulman), agents and publishers, and wonderful fellow writers (especially Jeff Hoffman) fostered the book's development.

Books reach readers with the help of great agents and publishers. We were fortunate to have both. We thank Gail Ross, Dara Kaye, and the remarkable group at Ross Yoon Agency for their patient and dedicated support of all of us, most of whom were first-time book authors, throughout the process. They took us to the excellent team at PublicAffairs, including Colleen Lawrie and Clive Priddle, who helped bring *Ending Parkinson's Disease* to fruition. They were supported by Miguel Cervantes, Lindsay Fradkoff, Brooke Parsons, Brynn Warriner, and colleagues.

We have also been helped by outstanding editors to make our academic prose more accessible to readers. Sara LaJeunesse and especially Susannah Meadows helped transform our writing. For six weeks, Susannah, who previously worked at *Newsweek* and authored *The Other Side of Impossible*, edited each sentence of the book at least twice. She pushed for clarity, deeper insights, journalistic standards, richer detail, and a readable text. We hope that we succeeded.

More generally, *Ending Parkinson's Disease* would not have been possible without the help and support of three communities. The first is those affected by Parkinson's. The impressive ways in which individuals and families cope with this debilitating disease is a never-ending source of education and inspiration. Collectively they are, in every respect, the muse that motivated us to write this book. While we have devoted much of our training and professional careers to the disease and related conditions, none of us has Parkinson's. We have not spent a day with a tremor or had our thinking slowed, voices softened, or independence eroded. Likewise, we have not served as care partners on a daily basis for those with the disease. This lack of personal experience profoundly limits our understanding and appreciation of its consequences. In this book, we hope to have provided a voice to those affected by Parkinson's.

The book also rests on the community of Parkinson's disease researchers, clinicians, and multidisciplinary care teams. We stand on the shoulders of scientific and medical giants and teachers who have characterized, identified the causes of, and developed treatments for Parkinson's. We also have benefited from clinical leaders from multiple disciplines who have developed better ways to meet the needs of people with Parkinson's. Their contributions, some of which are detailed in this book, have enabled us to identify better ways to address the disease. The work of these scientists and clinicians is often conducted over decades and done outside the public eye. Their contributions have shaped and formed the basis of many ideas in the book. Much of the credit for this book and our thanks go to them.

Finally, we thank our friends and families for encouraging us in this effort. We and our work are reflections of the environments that surround us. Our families have provided us the most honest and helpful critiques, allowed us to spend the necessary time to write, and supported us in countless ways. It is safe to say that they, too, have become Parkinson's fighters. We extend our immense appreciation and gratitude to all of them, especially to our spouses, Zena, Kelly, Leslie, and Inge. We love you.

Notes

INTRODUCTION

1. Dubos R. *Mirage of Health: Utopias, Progress, and Biological Change.* New York: Harper & Brothers; 1959.

2. Dorsey ER, Bloem BR. The Parkinson pandemic—a call to action. *JAMA Neurology.* 2018;75(1):9–10.

3. GBD 2015 Neurological Disorders Collaborator Group. Global, regional, and national burden of neurological disorders during 1990–2015: a systematic analysis for the Global Burden of Disease Study 2015. *Lancet Neurology.* 2017;16(11):877–897.

4. Dorsey ER, Bloem BR. The Parkinson pandemic—a call to action. *JAMA Neurology.* 2018;75(1):9–10.

5. Ibid.

6. Ascherio A, Schwarzschild MA. The epidemiology of Parkinson's disease: risk factors and prevention. *Lancet Neurology.* 2016;15(12):1257–1272; Goldman SM. Environmental toxins and Parkinson's disease. *Annual Review of Pharmacology and Toxicology.* 2014;54(1):141–164; Elbaz A, Carcaillon L, Kab S, Moisan F. Epidemiology of Parkinson's disease. *Revue Neurologique.* 2016;172(1):14–26.

7. Shulman LM, Katzel LI, Ivey FM, et al. Randomized clinical trial of 3 types of physical exercise for patients with Parkinson disease. *JAMA Neurology.* 2013;70(2):183–190; Rosenthal LS, Dorsey ER. The benefits of exercise

in Parkinson disease. *JAMA Neurology*. 2013;70(2):156–157; Li F, Harmer P, Fitzgerald K, et al. Tai chi and postural stability in patients with Parkinson's disease. *New England Journal of Medicine*. 2012;366(6):511–519.

8. Goldman SM. Environmental toxins and Parkinson's disease. *Annual Review of Pharmacology and Toxicology*. 2014;54(1):141–164; Deuschl G, Schade-Brittinger C, Krack P, et al. A randomized trial of deep-brain stimulation for Parkinson's disease. *New England Journal of Medicine*. 2006;355(9):896–908.

9. Factor SA, Feustel PJ, Friedman JH, et al. Longitudinal outcome of Parkinson's disease patients with psychosis. *Neurology*. 2003;60(11):1756–1761; Parashos SA, Maraganore DM, O'Brien PC, Rocca WA. Medical services utilization and prognosis in Parkinson disease: a population-based study. *Mayo Clinic Proceedings*. 2002;77(9):918–925; Schrag A, Hovris A, Morley D, Quinn N, Jahanshahi M. Caregiver-burden in Parkinson's disease is closely associated with psychiatric symptoms, falls, and disability. *Parkinsonism & Related Disorders*. 2006;12(1):35–41; Aarsland D, Larsen JP, Karlsen K, Lim NG, Tandberg E. Mental symptoms in Parkinson's disease are important contributors to caregiver distress. *International Journal of Geriatric Psychiatry*. 1999;14(10):866–874.

10. Buter TC, van den Hout A, Matthews FE, Larsen JP, Brayne C, Aarsland D. Dementia and survival in Parkinson disease: a 12-year population study. *Neurology*. 2008;70(13):1017–1022; Beyer MK, Herlofson K, Aarsland D, Larsen JP. Causes of death in a community-based study of Parkinson's disease. *Acta Neurologica Scandinavica*. 2001;103(1):7–11; Fall PA, Saleh A, Fredrickson M, Olsson J-E, Granérus A-K. Survival time, mortality, and cause of death in elderly patients with Parkinson's disease: a 9-year follow-up. *Movement Disorders*. 2003;18(11):1312–1316.

11. Parkinson J. An essay on the shaking palsy. *Journal of Neuropsychiatry and Clinical Neurosciences*. 2002;14(2):223–236.

12. The 20th century mortality files, 1901–2000. Office for National Statistics. March 11, 2011. https://webarchive.nationalarchives.gov.uk/20150908090558; www.ons.gov.uk/ons/publications/re-reference-tables.html?edition=tcm%3A77-215593. Accessed January 22, 2019; Duvoisin RC, Schweitzer MD. Paralysis agitans mortality in England and Wales, 1855–1962. *British Journal of Preventive and Social Medicine*. 1966;20(1):27–33.

13. More C. *Understanding the Industrial Revolution*. 1st ed. Routledge; 2000.

14. Ritchie H. What the history of London's air pollution can tell us about the future of today's growing megacities. Our World in Data. June 20, 2017. https://ourworldindata.org/london-air-pollution. Accessed Decem-

ber 12, 2018; Roser M, Ritchie H. Fertilizer and pesticides. Our World in Data. 2019. https://ourworldindata.org/fertilizer-and-pesticides. Accessed March 25, 2019; Manganese: history and industry development. Metalpedia. http://metalpedia.asianmetal.com/metal/manganese/history.shtml. Accessed March 28, 2019; Bell T. The history of steel: From Iron Age to electric arc furnaces. The Balance. www.thebalance.com/steel-history-2340172. Updated June 25, 2019. Accessed March 28, 2019; Caudle WM, Guillot TS, Lazo CR, Miller GW. Industrial toxicants and Parkinson's disease. Neurotoxicology, 2012;33(2):178–188.

15. Roser M, Ritchie H. Fertilizer and pesticides. Our World in Data. https://ourworldindata.org/fertilizer-and-pesticides. Accessed March 25, 2019.

16. Ascherio A, Chen H, Weisskopf MG, et al. Pesticide exposure and risk for Parkinson's disease. Annals of Neurology. 2006;60(2):197–203; Semchuk KM, Love EJ, Lee RG. Parkinson's disease and exposure to agricultural work and pesticide chemicals. Neurology. 1992;42(7):1328–1328; Pezzoli G, Cereda E. Exposure to pesticides or solvents and risk of Parkinson disease. Neurology. 2013;80(22):2035–2041; Tanner CM, Kamel F, Ross GW, et al. Rotenone, paraquat, and Parkinson's disease. Environmental Health Perspectives. 2011;119(6):866–872.

17. Caudle WM, Guillot TS, Lazo CR, Miller GW. Industrial toxicants and Parkinson's disease. Neurotoxicology. 2012;33(2):178–188; Betarbet R, Sherer TB, Greenamyre JT. Animal models of Parkinson's disease. BioEssays. 2002;24(4):308–318; Uversky VN. Neurotoxicant-induced animal models of Parkinson's disease: understanding the role of rotenone, maneb and paraquat in neurodegeneration. Cell and Tissue Research. 2004;318(1):225–241.

18. Hogue C. Trump administration delays bans of toxic solvents. Chemical & Engineering News. December 20, 2017. https://cen.acs.org/articles/95/web/2017/12/Trump-administration-delays-bans-toxic.html. Accessed March 28, 2019.

19. Barringer F. E.P.A. charts risks of a ubiquitous chemical. New York Times. September 30, 2011; Vettel P. Worried about decaf coffee? Fears are groundless. Chicago Tribune. May 19, 1985. www.chicagotribune.com/news/ct-xpm-1985-05-19-8502010068-story.html. Accessed July 19, 2019.

20. What is Superfund? US Environmental Protection Agency. www.epa.gov/superfund/what-superfund. Updated November 30, 2018. Accessed March 28, 2019.

21. Record of decision: Modock Road Springs/DLS Sand and Gravel, Inc. Site, Town of Victor, Ontario County, New York, Site Number 8-35-013. New York State Department of Environmental Conservation. January 2010. www

.dec.ny.gov/docs/remediation_hudson_pdf/835013rod.pdf. Accessed March 29, 2019.

22. Trichloroethylene. US Environmental Protection Agency. April 1992. www.epa.gov/sites/production/files/2016-09/documents/trichloroethylene .pdf. Updated January 2000. Accessed March 28, 2019.

23. Trichloroethylene (TCE) general fact sheet. In Agency for Toxic Substances and Disease Registry, US Department of Health and Human Services. www.gsa.gov/cdnstatic/Trichloroethylene_%28TCE%29_NEW_general _fact_sheet.pdf. Accessed March 29, 2019.

24. Risk management for trichloroethylene (TCE). US Environmental Protection Agency. www.epa.gov/assessing-and-managing-chemicals-under -tsca/risk-management-trichloroethylene-tce. Updated December 14, 2017. Accessed November 16, 2018; Integrated Risk Information System: trichloroethylene. US Environmental Protection Agency. https://cfpub.epa.gov/ncea /iris2/chemicalLanding.cfm?substance_nmbr=199. Updated July 28, 2017. Accessed March 28, 2019.

25. Watts M. Paraquat. Pesticide Action Network Asia & the Pacific. February 2011. www.panna.org/sites/default/files/Paraquat%20monograph %20final%202011-1.pdf. Accessed June 28, 2019; China about to ban all sales and use of paraquat by 2020 [press release]. CCM Data & Business Intelligence, July 31, 2017. www.cnchemicals.com/Press/89866-China%20 about%20to%20ban%20all%20sales%20and%20use%20of%20paraquat %20by%202020.html. Accessed June 28, 2019.

26. Tanner CM, Kamel F, Ross GW, et al. Rotenone, paraquat, and Parkinson's disease. *Environmental Health Perspectives*. 2011;119(6):866–872.

27. Pesticide National Synthesis Project: estimated annual agricultural pesticide use. US Geological Survey. https://water.usgs.gov/nawqa/pnsp/usage /maps/show_map.php?year=2015&map=PARAQUAT&hilo=H. Updated September 11, 2018. Accessed March 25, 2019.

28. Jenkins D. Court grants EPA new hearing on pesticide ban. *Capital Press*. February 12, 2019; O'Neill E. Will an appeals court make the EPA ban a pesticide linked to serious health risks? *The Salt*. February 26, 2019. www .npr.org/sections/thesalt/2019/02/26/698227414/will-an-appeals-court -make-the-epa-ban-a-pesticide-linked-to-serious-health-risk. Accessed June 28, 2019.

29. Friedman L. E.P.A. won't ban chlorpyrifos, pesticide tied to children's health problems. New York Times. July 18, 2019. www.nytimes.com /2019/07/18/climate/epa-chlorpyrifos-pesticide-ban.html?smid=nytcore -ios-share.

30. Ford D, Easton DF, Stratton M, et al. Genetic heterogeneity and penetrance analysis of the BRCA1 and BRCA2 genes in breast cancer families. *American Journal of Human Genetics*. 1998;62(3):676–689; Danaei G, Vander Hoorn S, Lopez AD, Murray CJL, Ezzati M. Causes of cancer in the world: comparative risk assessment of nine behavioural and environmental risk factors. *The Lancet*. 2005;366(9499):1784–1793; Brody JG, Moysich KB, Humblet O, Attfield KR, Beehler GP, Rudel RA. Environmental pollutants and breast cancer. *Cancer*. 2007;109(S12):2667–2711; McPherson K, Steel CM, Dixon JM. ABC of breast diseases. Breast cancer—epidemiology, risk factors, and genetics. *BMJ*. 2000;321(7261):624–628.

31. Hofman A, Stricker BH, Ikram MA, Koudstaal PJ, Darweesh SKL. Trends in the incidence of Parkinson disease in the general population: the Rotterdam Study. *American Journal of Epidemiology*. 2016;183(11):1018–1026.

32. Betarbet R, Sherer TB, MacKenzie G, Garcia-Osuna M, Panov AV, Greenamyre JT. Chronic systemic pesticide exposure reproduces features of Parkinson's disease. *Nature Neuroscience*. 2000;3(12):1301–1306.

33. Okun MS. *Parkinson's Treatment: 10 Secrets to a Happier Life*. CreateSpace; 2013; Okun MS, Gallo BV, Mandybur G, et al. Subthalamic deep brain stimulation with a constant-current device in Parkinson's disease: an open-label randomised controlled trial. *Lancet Neurology*. 2012;11(2):140–149; Okun MS, Fernandez HH, Wu SS, et al. Cognition and mood in Parkinson's disease in subthalamic nucleus versus globus pallidus interna deep brain stimulation: the COMPARE Trial. *Annals of Neurology*. 2009;65(5):586–595; Okun MS. *Ask the Doctor About Parkinson's Disease*. 1st ed. Demos Health; 2009. Okun MS. *10 Breakthrough Therapies for Parkinson's Disease: English Edition*. 1st ed. Books4Patients; 2015.

34. Bloem BR, Hausdorff JM, Visser JE, Giladi N. Falls and freezing of gait in Parkinson's disease: a review of two interconnected, episodic phenomena. *Movement Disorders*. 2004;19(8):871–884; Bloem BR, Grimbergen YAM, Cramer M, Willemsen M, Zwinderman AH. Prospective assessment of falls in Parkinson's disease. *Journal of Neurology*. 2001;248(11):950–958; Nijkrake MJ, Keus SHJ, Overeem S, et al. The ParkinsonNet concept: development, implementation and initial experience. *Movement Disorders*. 2010;25(7):823–829; Bloem BR, Rompen L, de Vries NM, Klink A, Munneke M, Jeurissen P. ParkinsonNet: a low-cost health care innovation with a systems approach from the Netherlands. *Health Affairs*. 2017;36:1987–1996.

35. Beck CA, Beran DB, Biglan KM, et al. National randomized controlled trial of virtual house calls for Parkinson disease. *Neurology*. 2017;89(11):1152–1161; Dorsey ER, Venkataraman V, Grana MJ, et al.

Randomized controlled clinical trial of "virtual house calls" for Parkinson disease. *JAMA Neurology.* 2013;70(5):565–570.

CHAPTER 1: SIX MEN IN LONDON

1. Parkinson J. An essay on the shaking palsy. *Journal of Neuropsychiatry and Clinical Neurosciences.* 2002;14(2):223–236.

2. Morris AD, Rose FC. *James Parkinson: His Life and Times.* Boston: Birkhauser; 1989.

3. Blake, W. The Poetical Works of William Blake, ed. by John Sampson. London, New York: Oxford University Press, 1908; Bartleby.com, 2011. www.bartleby.com/235/284.html.

4. Ritchie H. What the history of London's air pollution can tell us about the future of today's growing megacities. Our World in Data. June 20, 2017. https://ourworldindata.org/london-air-pollution. Accessed December 12, 2018.

5. Ibid.

6. Morris AD, Rose FC. *James Parkinson: His Life and Times.* Birkhauser; 1989.

7. Ibid.

8. Ibid.

9. Ibid.

10. Zhang Z, Dong Z, Román GC. Early descriptions of Parkinson disease in ancient China. *Archives of Neurology.* 2006;63(5):782–784; Ovallath S, Deepa P. The history of parkinsonism: descriptions in ancient Indian medical literature. *Movement Disorders.* 2013;28(5):566–568; Obeso JA, Stamelou M, Goetz CG, et al. Past, present, and future of Parkinson's disease: a special essay on the 200th anniversary of the shaking palsy. *Movement Disorders.* 2017;32(9):1264–1310; Lees AJ. Unresolved issues relating to the shaking palsy on the celebration of James Parkinson's 250th birthday. *Movement Disorders.* 2007;22(S17):S327–S334.

11. Zhang Z, Dong Z, Román GC. Early descriptions of Parkinson disease in ancient China. *Archives of Neurology.* 2006;63(5):782–784; Ovallath S, Deepa P. The history of parkinsonism: descriptions in ancient Indian medical literature. *Movement Disorders.* 2013;28(5):566–568.

12. Obeso JA, Stamelou M, Goetz CG, et al. Past, present, and future of Parkinson's disease: a special essay on the 200th anniversary of the shaking palsy. *Movement Disorders.* 2017;32(9):1264–1310.

13. Goetz CG. The history of Parkinson's disease: early clinical descriptions and neurological therapies. *Cold Spring Harbor Perspectives in Medicine.* 2011;1(1):a008862.

14. Obeso JA, Stamelou M, Goetz CG, et al. Past, present, and future of Parkinson's disease: a special essay on the 200th anniversary of the shaking palsy. *Movement Disorders.* 2017;32(9):1264–1310; Goetz CG. The history of Parkinson's disease: early clinical descriptions and neurological therapies. *Cold Spring Harbor Perspectives in Medicine.* 2011;1(1):a008862.

15. Osler W, Sir. *The Principles and Practice of Medicine.* New York: D. Appleton and Company; 1892.

16. Carlsson A. Nobel lecture: a half-century of neurotransmitter research: impact on neurology and psychiatry. *Bioscience Reports.* 2001;21(6):691–710.

17. Ibid.; Lees AJ, Tolosa E, Olanow CW. Four pioneers of L-dopa treatment: Arvid Carlsson, Oleh Hornykiewicz, George Cotzias, and Melvin Yahr. *Movement Disorders.* 2015;30(1):19–36.

18. Lees AJ, Tolosa E, Olanow CW. Four pioneers of L-dopa treatment: Arvid Carlsson, Oleh Hornykiewicz, George Cotzias, and Melvin Yahr. *Movement Disorders.* 2015;30(1):19–36.

19. Carlsson A. Nobel lecture: a half-century of neurotransmitter research: impact on neurology and psychiatry. *Bioscience Reports.* 2001;21(6):691–710.

20. Hagerty JR. Arvid Carlsson, snubbed at University in Sweden, found a path to Nobel Prize in medicine. *Wall Street Journal.* July 13, 2018.

21. Bernheimer H, Birkmayer W, Hornykiewicz O, Jellinger K, Seitelberger F. Brain dopamine and the syndromes of Parkinson and Huntington. Clinical, morphological and neurochemical correlations. *Journal of the Neurological Sciences.* 1973;20(4):415–455.

22. Lees AJ, Tolosa E, Olanow CW. Four pioneers of L-dopa treatment: Arvid Carlsson, Oleh Hornykiewicz, George Cotzias, and Melvin Yahr. *Movement Disorders.* 2015;30(1):19–36.

23. Braak H, Ghebremedhin E, Rüb U, Bratzke H, Del Tredici K. Stages in the development of Parkinson's disease–related pathology. *Cell and Tissue Research.* 2004;318(1):121–134; Braak H, Del Tredici K, Bratzke H, Hamm-Clement J, Sandmann-Keil D, Rüb U. Staging of the intracerebral inclusion body pathology associated with idiopathic Parkinson's disease (preclinical and clinical stages). *Journal of Neurology.* 2002;249(3):iii1–iii5.

24. Chaudhuri KR, Healy DG, Schapira AHV. Non-motor symptoms of Parkinson's disease: diagnosis and management. *Lancet Neurology.* 2006;5(3):235–245; Park A, Stacy M. Non-motor symptoms in Parkinson's disease. *Journal of Neurology.* 2009;256(3):293–298.

25. Schrag A. Quality of life and depression in Parkinson's disease. *Journal of the Neurological Sciences*. 2006;248(1):151–157; Shulman LM, Taback RL, Bean J, Weiner WJ. Comorbidity of the nonmotor symptoms of Parkinson's disease. *Movement Disorders*. 2001;16(3):507–510; Martinez-Martin P, Rodriguez-Blazquez C, Kurtis MM, Chaudhuri KR. The impact of non-motor symptoms on health-related quality of life of patients with Parkinson's disease. *Movement Disorders*. 2011;26(3):399–406.

26. Lees AJ, Tolosa E, Olanow CW. Four pioneers of L-dopa treatment: Arvid Carlsson, Oleh Hornykiewicz, George Cotzias, and Melvin Yahr. *Movement Disorders*. 2015;30(1):19–36.

27. Goetz CG. The history of Parkinson's disease: early clinical descriptions and neurological therapies. *Cold Spring Harbor Perspectives in Medicine*. 2011;1(1):a008862; Birkmayer W, Hornykiewicz O. The L-3,4-dioxyphenyl-alanine (DOPA)–effect in Parkinson-akinesia. *Wiener klinische Wochenschrift*. 1961;73:787–788.

28. Cotzias GC. L-dopa for Parkinsonism. *New England Journal of Medicine*. 1968;278(11):630; Cotzias GC, Papavasiliou PS, Gellene R. Modification of parkinsonism—chronic treatment with L-dopa. *New England Journal of Medicine*. 1969;280(7):337–345; Yahr MD, Duvoisin RC, Schear MJ, Barrett RE, Hoehn MM. Treatment of parkinsonism with levodopa. *Archives of Neurology*. 1969;21(4):343–354.

29. Sacks O. *Awakenings*. Rpt. ed. Vintage; 1999.

30. Parkinson J. An essay on the shaking palsy. *Journal of Neuropsychiatry and Clinical Neurosciences*. 2002;14(2):223–236.

31. Osler W, Sir. *The Principles and Practice of Medicine*. D. Appleton and Company; 1892.

32. Langston JW. The MPTP story. *Journal of Parkinson's Disease*. 2017;7(s1):S11–S19.

33. Ibid.

34. Langston JW, Palfreman J. *The Case of the Frozen Addicts: How the Solution of a Medical Mystery Revolutionized the Understanding of Parkinson's Disease*. Rev. ed. IOS Press; 2014.

35. Langston JW. The MPTP story. *Journal of Parkinson's Disease*. 2017;7(s1):S11–S19; Nonnekes J, Post B, Tetrud JW, Langston JW, Bloem BR. MPTP-induced parkinsonism: an historical case series. *Lancet Neurology*. 2018;17(4):300–301.

36. Langston JW, Ballard P, Tetrud JW, Irwin I. Chronic parkinsonism in humans due to a product of meperidine-analog synthesis. *Science*. 1983;219(4587):979–980.

37. "NOVA; case of the frozen addict, the." OpenVault from WGBH Boston/BBC TV. February 18, 1986. http://openvault.wgbh.org/catalog/V_474CF2C8A20B4173988486AC4C605A3C. At 00:55:25.

38. Cervone C. Frozen addicts; drug addicts developed Parkinson's disease. *Las Vegas World News*. December 9, 2013. www.lasvegasworldnews.com /frozen-addicts-drug-addicts-developed-parkinsons-disease/14976. Accessed December 13, 2018.

39. Desmethylprodine. Wikipedia. https://en.wikipedia.org/wiki /Desmethylprodine. Updated December 31, 2018. Accessed March 31, 2019.

40. Hughes AJ, Daniel SE, Kilford L, Lees AJ. Accuracy of clinical diagnosis of idiopathic Parkinson's disease: a clinico-pathological study of 100 cases. *Journal of Neurology, Neurosurgery, and Psychiatry*. 1992;55(3):181–184.

41. Davis GC, Williams AC, Markey SP, et al. Chronic parkinsonism secondary to intravenous injection of meperidine analogues. *Psychiatry Research*. 1979;1(3):249–254.

42. Ibid.

43. Designer drugs: hearing before the Committee on the Budget, United States Senate, Ninety-Ninth Congress, first session, July 18, 1985. US Government Publishing Office; 1985.

44. Ibid.

45. Sinclair M. Tragedy has designer label. *Washington Post*. July 19, 1985.

46. Ibid.

47. Langston JW, Palfreman J. *The Case of the Frozen Addicts: How the Solution of a Medical Mystery Revolutionized the Understanding of Parkinson's Disease*. Rev. ed. IOS Press; 2014.

48. Davis GC, Williams AC, Markey SP, et al. Chronic parkinsonism secondary to intravenous injection of meperidine analogues. *Psychiatry Research*. 1979;1(3):249–254.

49. Ibid.

50. Langston JW, Palfreman J. *The Case of the Frozen Addicts: How the Solution of a Medical Mystery Revolutionized the Understanding of Parkinson's Disease*. Rev. ed. IOS Press; 2014.

51. Designer drugs: hearing before the Committee on the Budget, United States Senate, Ninety-Ninth Congress, first session, July 18, 1985. US Government Publishing Office; 1985.

52. Sinclair M. Tragedy has designer label. *Washington Post*. July 19, 1985.

53. Davis GC, Williams AC, Markey SP, et al. Chronic parkinsonism secondary to intravenous injection of meperidine analogues. *Psychiatry Research*. 1979;1(3):249–254.

54. Designer drugs: hearing before the Committee on the Budget, United States Senate, Ninety-Ninth Congress, first session, July 18, 1985. US Government Publishing Office; 1985.

55. Langston JW. The MPTP story. *Journal of Parkinson's Disease.* 2017;7(s1):S11–S19.

56. Ibid.

57. Goldman SM. Environmental toxins and Parkinson's disease. *Annual Review of Pharmacology and Toxicology.* 2014;54(1):141–164; Betarbet R, Sherer TB, MacKenzie G, Garcia-Osuna M, Panov AV, Greenamyre JT. Chronic systemic pesticide exposure reproduces features of Parkinson's disease. *Nature Neuroscience.* 2000;3(12):1301–1306; Fleming L, Mann JB, Bean J, Briggle T, Sanchez-Ramos JR. Parkinson's disease and brain levels of organochlorine pesticides. *Annals of Neurology.* 1994;36(1):100–103; Priyadarshi A, Khuder SA, Schaub EA, Shrivastava S. A meta-analysis of Parkinson's disease and exposure to pesticides. *Neurotoxicology.* 2000;21(4):435–440.

58. Langston JW. The MPTP story. *Journal of Parkinson's Disease.* 2017;7(s1):S11–S19.

59. "NOVA; case of the frozen addict, the." OpenVault from WGBH Boston/BBC TV. February 18, 1986. http://openvault.wgbh.org/catalog /V_474CF2C8A20B4173988486AC4C605A3C. At 00:55:25.

60. Meredith GE, Rademacher DJ. MPTP mouse models of Parkinson's disease: an update. *Journal of Parkinson's Disease.* 2011;1(1):19–33; Porras G, Li Q, Bezard E. Modeling Parkinson's disease in primates: the MPTP model. *Cold Spring Harbor Perspectives in Medicine.* 2012;2(3):a009308.

61. Cornelius CE. Animal models—a neglected medical resource. *New England Journal of Medicine.* 1969;281(17):934–944; Overall KL. Natural animal models of human psychiatric conditions: assessment of mechanism and validity. *Progress in Neuro-psychopharmacology and Biological Psychiatry.* 2000;24(5):727–776.

62. Barré-Sinoussi F, Montagutelli X. Animal models are essential to biological research: issues and perspectives. *Future Science OA.* 2015;1(4):FSO63–FSO63.

63. Ibid.

64. Giasson BI, Lee VMY. A new link between pesticides and Parkinson's disease. *Nature Neuroscience.* 2000;3:1227–1228.

65. Betarbet R, Sherer TB, MacKenzie G, Garcia-Osuna M, Panov AV, Greenamyre JT. Chronic systemic pesticide exposure reproduces features of Parkinson's disease. *Nature Neuroscience.* 2000;3(12):1301–1306.

66. Giasson BI, Lee VMY. A new link between pesticides and Parkinson's disease. *Nature Neuroscience.* 2000;3:1227–1228.

67. Tanner CM, Kamel F, Ross GW, et al. Rotenone, paraquat, and Parkinson's disease. *Environmental Health Perspectives.* 2011;119(6):866–872.

68. Polymeropoulos MH, Lavedan C, Leroy E, et al. Mutation in the α-synuclein gene identified in families with Parkinson's disease. *Science.* 1997;276(5321):2045–2047.

69. Olanow CW, Brundin P. Parkinson's disease and alpha synuclein: is Parkinson's disease a prion-like disorder? *Movement Disorders.* 2013;28(1):31–40; Uversky VN, Li J, Fink AL. Pesticides directly accelerate the rate of α-synuclein fibril formation: a possible factor in Parkinson's disease. *FEBS Letters.* 2001;500(3):105–108.

70. Luk KC, Kehm V, Carroll J, et al. Pathological α-synuclein transmission initiates Parkinson-like neurodegeneration in nontransgenic mice. *Science.* 2012;338(6109):949–953.

71. Goedert M, Spillantini MG, Del Tredici K, Braak H. 100 years of Lewy pathology. *Nature Reviews Neurology.* 2013;9(1):13–24.

72. Spillantini MG, Schmidt ML, Lee VMY, Trojanowski JQ, Jakes R, Goedert M. α-synuclein in Lewy bodies. *Nature.* 1997;388:839–840; Mezey E, Dehejia AM, Harta G, et al. Alpha synuclein is present in Lewy bodies in sporadic Parkinson's disease. *Molecular Psychiatry.* 1998;3:493–499.

73. Holdorff B. Friedrich Heinrich Lewy (1885–1950) and his work. *Journal of the History of the Neurosciences.* 2002;11(1):19–28.

74. Brin S. LRRK2. *Too.* September 18, 2008. http://too.blogspot.com/2008/09/lrrk2.html.

75. Zimprich A, Biskup S, Leitner P, et al. Mutations in LRRK2 cause autosomal-dominant parkinsonism with pleomorphic pathology. *Neuron.* 2004;44(4):601–607.

76. Ibid.; Gilks WP, Abou-Sleiman PM, Gandhi S, et al. A common LRRK2 mutation in idiopathic Parkinson's disease. *The Lancet.* 2005;365(9457):415–416; Di Fonzo A, Rohé CF, Ferreira J, et al. A frequent LRRK2 gene mutation associated with autosomal dominant Parkinson's disease. *The Lancet.* 2005;365(9457):412–415.

77. Brin S. LRRK2. *Too.* September 18, 2008. http://too.blogspot.com/2008/09/lrrk2.html.

78. Ibid.

79. Sergey Brin. Inside Philanthropy. www.insidephilanthropy.com/guide-to-individual-donors/sergey-brin.html. Accessed March 20, 2019; Goetz T. Sergey Brin's search for a Parkinson's cure. *Wired.* January 22, 2010. www.wired.com/2010/06/ff-sergeys-search. Accessed June 28, 2019.

80. The Associated Press. Twins study links Parkinson's disease to the environment. *New York Times.* February 2, 1999.

81. Klein C, Westenberger A. Genetics of Parkinson's disease. *Cold Spring Harbor Perspectives in Medicine*. 2012;2(1):a008888; Fleming SM. Mechanisms of gene-environment interactions in Parkinson's disease. *Current Environmental Health Reports*. 2017;4(2):192–199.

82. Klein C, Westenberger A. Genetics of Parkinson's disease. *Cold Spring Harbor Perspectives in Medicine*. 2012;2(1):a008888.

83. Fleming SM. Mechanisms of gene-environment interactions in Parkinson's disease. *Current Environmental Health Reports*. 2017;4(2):192–199.

84. www.reuters.com/article/us-cancer-lung-nutrients-sb/nutrients-may-be -why-some-smokers-avoid-cancer-idUSTRE65E5JW20100616.

85. www.nature.com/articles/nature06846; www.nature.com/articles/ng .3892.

86. www.ncbi.nlm.nih.gov/pmc/articles/PMC4501942; https://pubmed .ncbi.nlm.nih.gov/28639421-penetrance-estimate-of-lrrk2-pg2019s -mutation-in-individuals-of-non-ashkenazi-jewish-ancestry.

87. Ryan SD, Dolatabadi N, Chan SF, et al. Isogenic human iPSC Parkinson's model shows nitrosative stress-induced dysfunction in MEF2-PGC1α transcription. *Cell*. 2013;155(6):1351–1364.

88. Nuber S, Tadros D, Fields J, et al. Environmental neurotoxic challenge of conditional alpha-synuclein transgenic mice predicts a dopaminergic olfactory-striatal interplay in early PD. *Acta Neuropathologica*. 2014;127(4):477–494.

89. Liu H-F, Ho PW-L, Leung GC-T, et al. Combined LRRK2 mutation, aging and chronic low dose oral rotenone as a model of Parkinson's disease. *Scientific Reports*. 2017;7:40887. doi:10.1038/srep40887.

90. Weiner WJ. There is no Parkinson disease. *Archives of Neurology*. 2008;65(6):705–708.

91. Malek N, Weil RS, Bresner C, et al. Features of GBA-associated Parkinson's disease at presentation in the UK Tracking Parkinson's study. *Journal of Neurology, Neurosurgery & Psychiatry*. 2018;89(7):702–709; Saunders-Pullman R, Mirelman A, Alcalay RN, et al. Progression in the LRRK2-associated Parkinson disease population. *JAMA Neurology*. 2018;75(3):312–319.

92. Parkinson J. An essay on the shaking palsy. *Journal of Neuropsychiatry and Clinical Neurosciences*. 2002;14(2):223–236.

93. Braak H, Del Tredici K, Rüb U, de Vos RAI, Jansen Steur ENH, Braak E. Staging of brain pathology related to sporadic Parkinson's disease. *Neurobiology of Aging*. 2003;24(2):197–211; Braak H, Rüb U, Gai WP, Del Tredici K. Idiopathic Parkinson's disease: possible routes by which vulnerable neuronal types may be subject to neuroinvasion by an unknown pathogen. *Journal of Neural Transmission*. 2003;110(5):517–536.

94. Braak H, Rüb U, Gai WP, Del Tredici K. Idiopathic Parkinson's disease: possible routes by which vulnerable neuronal types may be subject to neuroinvasion by an unknown pathogen. *Journal of Neural Transmission.* 2003;110(5):517–536.

95. Silva BA, Einarsdóttir O, Fink AL, Uversky VN. Biophysical characterization of α-synuclein and rotenone interaction. *Biomolecules.* 2013;3(3):703–732.

96. Del Tredici K, Hawkes CH, Ghebremedhin E, Braak H. Lewy pathology in the submandibular gland of individuals with incidental Lewy body disease and sporadic Parkinson's disease. *Acta Neuropathologica.* 2010;119(6):703–713; Del Tredici K, Braak H. Spinal cord lesions in sporadic Parkinson's disease. *Acta Neuropathologica.* 2012;124(5):643–664.

97. Breen DP, Halliday GM, Lang AE. Gut-brain axis and the spread of α-synuclein pathology: vagal highway or dead end? *Movement Disorders.* 2019;34(3):307–316.

98. Braak H, Rüb U, Gai WP, Del Tredici K. Idiopathic Parkinson's disease: possible routes by which vulnerable neuronal types may be subject to neuroinvasion by an unknown pathogen. *Journal of Neural Transmission.* 2003;110(5):517–536.

99. Pan-Montojo F, Anichtchik O, Dening Y, et al. Progression of Parkinson's disease pathology is reproduced by intragastric administration of rotenone in mice. *PLOS ONE.* 2010;5(1):e8762; Holmqvist S, Chutna O, Bousset L, et al. Direct evidence of Parkinson pathology spread from the gastrointestinal tract to the brain in rats. *Acta Neuropathologica.* 2014;128(6):805–820.

100. Braak H, Del Tredici K, Rüb U, de Vos RAI, Jansen Steur ENH, Braak E. Staging of brain pathology related to sporadic Parkinson's disease. *Neurobiology of Aging.* 2003;24(2):197–211.

101. Stokholm MG, Danielsen EH, Hamilton-Dutoit SJ, Borghammer P. Pathological α-synuclein in gastrointestinal tissues from prodromal Parkinson disease patients. *Annals of Neurology.* 2016;79(6):940–949; Ross GW, Abbott RD, Petrovitch H, et al. Association of olfactory dysfunction with incidental Lewy bodies. *Movement Disorders.* 2006;21(12):2062–2067.

102. Chaudhuri KR, Healy DG, Schapira AHV. Non-motor symptoms of Parkinson's disease: diagnosis and management. *Lancet Neurology.* 2006;5(3):235–245; Winge K, Skau A-M, Stimpel H, Nielsen KK, Werdelin L. Prevalence of bladder dysfunction in Parkinsons disease. *Neurourology and Urodynamics.* 2006;25(2):116–122; Aarsland D, Andersen K, Larsen JP, Lolk A. Prevalence and characteristics of dementia in Parkinson disease: an 8-year prospective study. *Archives of Neurology.* 2003;60(3):387–392; Hely MA, Reid WGJ, Adena MA, Halliday GM, Morris JGL. The

Sydney multicenter study of Parkinson's disease: the inevitability of dementia at 20 years. *Movement Disorders.* 2008;23(6):837–844.

103. Svensson E, Horváth-Puhó E, Thomsen RW, et al. Vagotomy and subsequent risk of Parkinson's disease. *Annals of Neurology.* 2015;78(4):522–529.

CHAPTER 2: A MAN-MADE PANDEMIC

1. Parkinson's: an industrial age disease [transcript]. Living on Earth. August 15, 1997. www.loe.org/shows/segments.html?programID=97-P13-00033 &segmentID=1. Accessed March 26, 2019.

2. Poskanzer DC, Schwab RS. Studies in the epidemiology of Parkinson's disease predicting its disappearance as a major clinical entity by 1980. *Transactions of the American Neurological Association.* 1961;86:234–235.

3. Constantin von Economo. Wikipedia. https://en.wikipedia.org/wiki /Constantin_von_Economo. Accessed February 15, 2019; Sak J, Grzybowski A. Brain and aviation: on the 80th anniversary of Constantin von Economo's (1876–1931) death. *Neurological Sciences.* 2013;34(3):387–391.

4. Dickman MS. von Economo encephalitis. *Archives of Neurology.* 2001;58(10):1696–1698.

5. Lutters B, Foley P, Koehler PJ. The centennial lesson of encephalitis lethargica. *Neurology.* 2018;90(12):563–567.

6. Vilensky JA, Hoffman LA. Encephalitis lethargica: 100 years after the epidemic. *Brain.* 2017;140(8):2246–2251.

7. Sak J, Grzybowski A. Brain and aviation: on the 80th anniversary of Constantin von Economo's (1876–1931) death. *Neurological Sciences.* 2013;34(3):387–391; Dickman MS. von Economo encephalitis. *Archives of Neurology.* 2001;58(10):1696–1698; Estupinan D, Nathoo S, Okun MS. The demise of Poskanzer and Schwab's influenza theory on the pathogenesis of Parkinson's disease. *Parkinson's Disease.* 2013;2013:167843.

8. Lutters B, Foley P, Koehler PJ. The centennial lesson of encephalitis lethargica. *Neurology.* 2018;90(12):563–567.

9. Sacks O. *Awakenings.* Rpt. ed. Vintage; 1999.

10. Ibid.

11. Ibid.

12. Ibid.

13. Poskanzer DC, Schwab RS. Studies in the epidemiology of Parkinson's disease predicting its disappearance as a major clinical entity by 1980. *Transactions of the American Neurological Association.* 1961;86:234–235.

14. The Parkinson's puzzle. *Time.* 1974;104:71.

15. Moore G. Influenza and Parkinson's disease. *Public Health Reports.* 1977;92(1):79–80.

16. Hoehn MM. Age distribution of patients with Parkinsonism. *Journal of the American Geriatrics Society.* 1976;24(2):79–85.

17. Margaret "Peggy" Hoehn M.D. *Denver Post.* July 22, 2005;Obituary.

18. GBD 2016 Parkinson's Disease Collaborators. Global, regional, and national burden of Parkinson's disease, 1990–2016: a systematic analysis for the Global Burden of Disease Study 2016. *Lancet Neurology.* 2018;17(11):939–953.

19. Muangpaisan W, Hori H, Brayne C. Systematic review of the prevalence and incidence of Parkinson's disease in Asia. *Journal of Epidemiology.* 2009;19(6):281–293.

20. Goldman SM. Environmental toxins and Parkinson's disease. *Annual Review of Pharmacology and Toxicology.* 2014;54(1):141–164.

21. Tilman D, Cassman KG, Matson PA, Naylor R, Polasky S. Agricultural sustainability and intensive production practices. *Nature.* 2002;418:671–677.

22. FAOSTAT: pesticides. Food and Agriculture Organization of the United Nations. www.fao.org/faostat/en/#data/EP/visualize. Accessed February 15, 2019.

23. Ibid.

24. Tilman D, Cassman KG, Matson PA, Naylor R, Polasky S. Agricultural sustainability and intensive production practices. *Nature.* 2002;418:671–677.

25. Rutchik JS. Organic solvent neurotoxicity. Medscape. https://emedicine.medscape.com/article/1174981-overview. Updated December 11, 2018. Accessed February 15, 2019.

26. Solvents. ChemicalSafetyFacts.org. www.chemicalsafetyfacts.org/solvents. Accessed March 26, 2019.

27. Dick FD. Solvent neurotoxicity. *Occupational and Environmental Medicine.* 2006;63(3):221–226.

28. C2 chlorinated solvents. IHS Markit. October 2017. https://ihsmarkit.com/products/c2-chlorinated-chemical-economics-handbook.html. Accessed March 26, 2019.

29. Ritz B, Lee P-C, Hansen J, et al. Traffic-related air pollution and Parkinson's disease in Denmark: a case-control study. *Environmental Health Perspectives.* 2016;124(3):351–356; Doty RL. Olfactory dysfunction in Parkinson disease. *Nature Reviews Neurology.* 2012;8:329–339; Calderón-Garcidueñas L, Azzarelli B, Acuna H, et al. Air pollution and brain damage. *Toxicologic Pathology.* 2002;30(3):373–389.

30. GBD 2016 Parkinson's Disease Collaborators. Global, regional, and national burden of Parkinson's disease, 1990–2016: a systematic

analysis for the Global Burden of Disease Study 2016. *Lancet Neurology*. 2018;17(11):939–953.

31. Rocheleau M. Chart: the percentage of women and men in each profession. Boston Globe. March 6, 2017. www.bostonglobe.com/metro/2017 /03/06/chart-the-percentage-women-and-men-each-profession/GBX22Ys WI0XaeHghwXfE4H/story.html. Accessed February 15, 2019.

32. GBD 2016 Parkinson's Disease Collaborators. Global, regional, and national burden of Parkinson's disease, 1990–2016: a systematic analysis for the Global Burden of Disease Study 2016. *Lancet Neurology*. 2018;17(11):939–953; Van Den Eeden SK, Tanner CM, Bernstein AL, et al. Incidence of Parkinson's disease: variation by age, gender, and race/ethnicity. *American Journal of Epidemiology*. 2003;157(11):1015–1022; Wooten GF, Currie LJ, Bovbjerg VE, Lee JK, Patrie J. Are men at greater risk for Parkinson's disease than women? *Journal of Neurology, Neurosurgery & Psychiatry*. 2004;75(4):637–639; Mayeux R, Marder K, Cote LJ, et al. The frequency of idiopathic Parkinson's disease by age, ethnic group, and sex in northern Manhattan, 1988–1993. *American Journal of Epidemiology*. 1995;142(8):820–827.

33. Roser M. Life expectancy. Our World in Data. https://ourworldindata .org/life-expectancy. Accessed March 31, 2019.

34. Ibid.

35. Paid notice: Deaths: Hoehn, Professor Margaret "Peggy," M.D. *New York Times*. July 24, 2005: 1001032.

36. International programs: International Data Base. US Census Bureau. www.census.gov/data-tools/demo/idb/informationGateway.php. Updated: September 2018. Accessed November 19, 2018; Population ages 65 and above, total. World Bank. https://data.worldbank.org/indicator/SP.POP.65UP.To. Accessed November 19, 2018.

37. Levy G. The relationship of Parkinson disease with aging. *Archives of Neurology*. 2007;64(9):1242–1246.

38. Savica R, Rocca WA, Ahlskog JE. When does Parkinson disease start? *Archives of Neurology*. 2010;67(7):798–801.

39. Van Den Eeden SK, Tanner CM, Bernstein AL, et al. Incidence of Parkinson's disease: variation by age, gender, and race/ethnicity. American Journal of Epidemiology. 2003;157(11):1015–1022. www.ncbi.nlm.nih.gov /pubmed/12777365.

40. Scutti S. Drug overdoses, suicides cause drop in 2017 US life expectancy; CDC director calls it a "wakeup call." *CNN*. December 17, 2018. www .cnn.com/2018/11/29/health/life-expectancy-2017-cdc/index.html. Accessed February 15, 2019; Naik G. Global life expectancy increases by about

six years: study in Lancet says rise is results of dramatic health-care advances. *Wall Street Journal*. December 18, 2014; Murray CJL, Barber RM, Foreman KJ, et al. Global, regional, and national disability-adjusted life years (DALYs) for 306 diseases and injuries and healthy life expectancy (HALE) for 188 countries, 1990–2013: quantifying the epidemiological transition. *The Lancet*. 2015;386(10009):2145–2191.

41. Wanneveich M, Moisan F, Jacqmin-Gadda H, Elbaz A, Joly P. Projections of prevalence, lifetime risk, and life expectancy of Parkinson's disease (2010–2030) in France. *Movement Disorders*. 2018;33(9):1449–1455.

42. Proctor RN. The history of the discovery of the cigarette–lung cancer link: evidentiary traditions, corporate denial, global toll. *Tobacco Control*. 2012;21(2):87–91.

43. Ibid.

44. Ibid.; Lung cancer. Wikipedia. https://en.wikipedia.org/wiki/Lung _cancer. Accessed February 15, 2019.

45. Keating C. *Smoking Kills: The Revolutionary Life of Richard Doll*. Oxford: Signal Books; 2009.

46. Doll R, Hill AB. Lung cancer and other causes of death in relation to smoking; a second report on the mortality of British doctors. *BMJ*. 1956;2(5001):1071–1081.

47. Proctor RN. The history of the discovery of the cigarette–lung cancer link: evidentiary traditions, corporate denial, global toll. *Tobacco Control*. 2012;21(2):87–91.

48. Flanders WD, Lally CA, Zhu B-P, Henley SJ, Thun MJ. Lung cancer mortality in relation to age, duration of smoking, and daily cigarette consumption. *Cancer Research*. 2003;63(19):6556–6562.

49. Pesch B, Kendzia B, Gustavsson P, et al. Cigarette smoking and lung cancer—relative risk estimates for the major histological types from a pooled analysis of case-control studies. *International Journal of Cancer*. 2012;131(5):1210–1219.

50. GBD 2016 Parkinson's Disease Collaborators. Global, regional, and national burden of Parkinson's disease, 1990–2016: a systematic analysis for the Global Burden of Disease Study 2016. *Lancet Neurology*. 2018;17(11):939–953; Li X, Li W, Liu G, Shen X, Tang Y. Association between cigarette smoking and Parkinson's disease: a meta-analysis. *Archives of Gerontology and Geriatrics*. 2015;61(3):510–516.

51. Ma C, Liu Y, Neumann S, Gao X. Nicotine from cigarette smoking and diet and Parkinson disease: a review. *Translational Neurodegeneration*. 2017;6:18; Maggio R, Riva M, Vaglini F, et al. Nicotine prevents experimental

parkinsonism in rodents and induces striatal increase of neurotrophic factors. *Journal of Neurochemistry*. 1998;71(6):2439–2446; Allam MF, Serrano del Castillo A, Fernandez-Crehuet Navajas R. Smoking and Parkinson's disease: explanatory hypothesis. *International Journal of Neuroscience*. 2002;112(7):851–854; Miksys S, Tyndale RF. Nicotine induces brain CYP enzymes: relevance to Parkinson's disease. *Journal of Neural Transmission, Supplement*. 2006(70):177–180.

52. Biedermann L, Zeitz J, Mwinyi J, et al. Smoking cessation induces profound changes in the composition of the intestinal microbiota in humans. *PLOS ONE*. 2013;8(3):e59260; Scheperjans F, Pekkonen E, Kaakkola S, Auvinen P. Linking smoking, coffee, urate, and Parkinson's disease—a role for gut microbiota? *Journal of Parkinson's Disease*. 2015;5(2):255–262.

53. Pagliuca G, Rosato C, Martellucci S, et al. Cytologic and functional alterations of nasal mucosa in smokers: temporary or permanent damage? *Otolaryngology—Head and Neck Surgery*. 2015;152(4):740–745; Utiyama DMO, Yoshida CT, Goto DM, et al. The effects of smoking and smoking cessation on nasal mucociliary clearance, mucus properties and inflammation. *Clinics*. 2016;71(6):344–350; Martin EM, Clapp PW, Rebuli ME, et al. E-cigarette use results in suppression of immune and inflammatory-response genes in nasal epithelial cells similar to cigarette smoke. *American Journal of Physiology—Lung Cellular and Molecular Physiology*. 2016;311(1):L135–L144.

54. Scheperjans F, Pekkonen E, Kaakkola S, Auvinen P. Linking smoking, coffee, urate, and Parkinson's disease—a role for gut microbiota? *Journal of Parkinson's Disease*. 2015;5(2):255–262; Derkinderen P, Shannon KM, Brundin P. Gut feelings about smoking and coffee in Parkinson's disease. *Movement Disorders*. 2014;29(8):976–979.

55. Hopfner F, Künstner A, Müller SH, et al. Gut microbiota in Parkinson disease in a northern German cohort. *Brain Research*. 2017;1667:41–45; Tremlett H, Bauer KC, Appel-Cresswell S, Finlay BB, Waubant E. The gut microbiome in human neurological disease: a review. *Annals of Neurology*. 2017;81(3):369–382; Hasegawa S, Goto S, Tsuji H, et al. Intestinal dysbiosis and lowered serum lipopolysaccharide-binding protein in Parkinson's disease. *PLOS ONE*. 2015;10(11):e0142164; Keshavarzian A, Green SJ, Engen PA, et al. Colonic bacterial composition in Parkinson's disease. *Movement Disorders*. 2015;30(10):1351–1360.

56. Scheperjans F, Pekkonen E, Kaakkola S, Auvinen P. Linking Smoking, Coffee, Urate, and Parkinson's Disease—A Role for Gut Microbiota? *Journal of Parkinson's Disease*. 2015;5(2):255–262.

57. Sampson TR, Debelius JW, Thron T, et al. Gut microbiota regulate motor deficits and neuroinflammation in a model of Parkinson's disease. *Cell.* 2016;167(6):1469–1480.e12.

58. Ibid.

59. Health risks of smoking tobacco. American Cancer Society. Cancer A–Z. www.cancer.org/cancer/cancer-causes/tobacco-and-cancer/health-risks -of-smoking-tobacco.html. Accessed February 15, 2019.

60. Dorsey ER, Sherer T, Okun M, Bloem BR. The emerging evidence of the Parkinson pandemic. *Journal of Parkinson's Disease.* 2018;8(Suppl 1):S3–S8.

61. Morris AD, Rose FC. *James Parkinson: His Life and Times.* Birkhauser; 1989.

62. Omran AR. The epidemiologic transition: a theory of the epidemiology of population change. *Milbank Quarterly.* 2005;83(4):731–757.

63. Fauci AS, Morens DM, Folkers GK. What is a Pandemic? *Journal of Infectious Diseases.* 2009;200(7):1018–1021.

64. Black death. Wikipedia. https://en.wikipedia.org/wiki/Black_Death. Accessed February 15, 2019.

65. Barry JM. *The Great Influenza: The Story of the Deadliest Pandemic in History.* Rev. ed: Penguin Books; 2005.

66. Lozano R, Naghavi M, Foreman K, et al. Global and regional mortality from 235 causes of death for 20 age groups in 1990 and 2010: a systematic analysis for the Global Burden of Disease Study 2010. *The Lancet.* 2012;380(9859):2095–2128.

67. Omran AR. The epidemiologic transition: a theory of the epidemiology of population change. *Milbank Quarterly.* 2005;83(4):731–757.

68. Olshansky SJ, Ault AB. The fourth stage of the epidemiologic transition: the age of delayed degenerative diseases. *Milbank Quarterly.* 1986;64(3):355–391.

69. Allen L. Are we facing a noncommunicable disease pandemic? *Journal of Epidemiology and Global Health.* 2017;7(1):5–9.

70. Fauci AS, Morens DM, Folkers GK. What is a Pandemic? *Journal of Infectious Diseases.* 2009;200(7):1018–1021.

71. GBD 2016 Parkinson's Disease Collaborators. Global, regional, and national burden of Parkinson's disease, 1990–2016: a systematic analysis for the Global Burden of Disease Study 2016. *Lancet Neurology.* 2018;17(11):939–953.

72. Ibid.

73. Gage H, Hendricks A, Zhang S, Kazis L. The relative health related quality of life of veterans with Parkinson's disease. *Journal of Neurology, Neurosurgery, and Psychiatry.* 2003;74(2):163–169.

74. Kochanek KD, Murphy SL, Xu J, Tejada-Vera B. *Deaths: Final Data for 2014.* US Department of Health and Human Services; June 30, 2016.

75. FAOSTAT: pesticides. Food and Agriculture Organization of the United Nations. www.fao.org/faostat/en/#data/EP/visualize. Accessed February 15, 2019; Friesen MC, Locke SJ, Chen Y-C, et al. Historical occupational trichloroethylene air concentrations based on inspection measurements from Shanghai, China. *Annals of Occupational Hygiene.* 2015;59(1):62–78; Tang Z. The characteristics of urban air pollution in China. In: *Urbanization, Energy, and Air Pollution in China: The Challenges Ahead: Proceedings of a Symposium.* National Academies Press; 2004:47–54.

76. GBD 2016 Parkinson's Disease Collaborators. Global, regional, and national burden of Parkinson's disease, 1990–2016: a systematic analysis for the Global Burden of Disease Study 2016. *Lancet Neurology.* 2018;17(11):939–953.

77. Parashos SA, Maraganore DM, O'Brien PC, Rocca WA. Medical services utilization and prognosis in Parkinson disease: a population-based study. *Mayo Clinic Proceedings.* 2002;77(9):918–925; Gage H, Hendricks A, Zhang S, Kazis L. The relative health related quality of life of veterans with Parkinson's disease. *Journal of Neurology, Neurosurgery, and Psychiatry.* 2003;74(2):163–169; Aarsland D, Larsen JP, Tandberg E, Laake K. Predictors of nursing home placement in Parkinson's disease: a population-based, prospective study. *Journal of the American Geriatrics Society.* 2000;48(8):938–942; Dorsey ER, Deuel LM, Voss TS, et al. Increasing access to specialty care: a pilot, randomized controlled trial of telemedicine for Parkinson's disease. *Movement Disorders.* 2010;25(11):1652–1659.

78. Hinz M, Stein A, Cole T. The Parkinson's disease death rate: carbidopa and vitamin B6. *Journal of Clinical Pharmacology.* 2014;6:161–169.

CHAPTER 3: VANQUISHING INDIFFERENCE

1. Furnish D. To those who stood against AIDS apathy: thank you. *Huffington Post.* November 8, 2013.

2. Litsky F, Weber B. Roger Bannister, first athlete to break the 4-minute mile, is dead at 88. *New York Times.* March 5, 2018; Ingle S. Interview: Roger Bannister turns clock back 60 years and still feels the thrill. *The Running Blog. Guardian.* May 5, 2014; Crouch I. Roger Bannister's solitary pursuit of the four-minute mile. *New Yorker.* March 5, 2018.

3. Guinness World Records. First sub–four minute mile—Sir Roger Bannister—Guinness World Records 60th anniversary [video file]. YouTube. December 8, 2014. www.youtube.com/watch?v=ku1vdrWQuJY.

4. Ibid.

5. Bannister R. Roger Bannister: "The day I broke the four-minute mile." *Daily Telegraph*. March 30, 2014.

6. Litsky F, Weber B. Roger Bannister, first athlete to break the 4-minute mile, is dead at 88. *New York Times*. March 5, 2018; Roger Bannister. In: Dana R. Barnes, ed. *Notable Sports Figures*. Vol 1. Gale Group; 2004:86.

7. Limb M. Roger Gilbert Bannister: innovative neurologist and the first athlete to run a mile in less than four minutes. *BMJ*. 2018;361:k1589

8. Litsky F, Weber B. Roger Bannister, first athlete to break the 4-minute mile, is dead at 88. *New York Times*. March 5, 2018.

9. Oshinsky DM. *Polio: An American Story*. Oxford University Press; 2006.

10. Ibid.

11. Ibid.

12. Goldman AS, Schmalstieg EJ, Freeman DH, Goldman DA, Schmalstieg FC. What was the cause of Franklin Delano Roosevelt's paralytic illness? *Journal of Medical Biography*. 2003;11(4):232–240.

13. Oshinsky DM. *Polio: An American Story*. Oxford University Press; 2006.

14. Ibid.

15. Ibid.

16. Ibid.

17. Ibid.

18. Ibid.

19. Ibid.

20. Ibid.

21. Admin. History of health care. *What If America Had a Healthcare System That Worked?* December 20, 2009. http://whatifpost.com/health-care-resources/history-of-health-care.

22. Oshinsky DM. *Polio: An American Story*. Oxford University Press; 2006.

23. Moore W. Paralysed with fear: the story of polio by Gareth Williams—review. *Guardian*. July 17, 2013;Culture.

24. Poliomyelitis. Wikipedia. https://en.wikipedia.org/wiki/Poliomyelitis. Accessed October 15, 2018.

25. Resnick B. What America looked like: polio children paralyzed in iron lungs. *Atlantic*. January 10, 2012.

26. Press TA. Woman dies after life spent in iron lung. *CNN*. 2008. https://web.archive.org/web/20081022162511/http:/www.cnn.com/2008/US/05/28/iron.lung.death.ap/index.html. Updated May 28. Accessed March 4, 2019.

27. Meldrum M. "A calculated risk": the Salk polio vaccine field trials of 1954. *BMJ*. 1998;317(7167):1233–1236.

28. Global Citizen. Could you patent the sun? [video file]. YouTube. January 29, 2013. www.youtube.com/watch?v=erHXKP386Nk.

29. Polio vaccine. Wikipedia. https://en.wikipedia.org/wiki/Polio_vaccine. Accessed October 15, 2018.

30. Oshinsky DM. *Polio: An American Story*. Oxford University Press; 2006.

31. Altman LK. Rare cancer seen in 41 homosexuals. *New York Times*. July 3, 1981:A00020.

32. Gupta RK, Abdul-Jawad S, McCoy LE, et al. HIV-1 remission following CCR5Δ32/Δ32 haematopoietic stem-cell transplantation. *Nature*. 2019;568(7751):244–248.

33. Roser M, Ritchie H. HIV/AIDS. Our World in Data. November 2014. https://ourworldindata.org/hiv-aids. Updated April 2018. Accessed March 31, 2019.

34. Lopez G. The Reagan administration's unbelievable response to the HIV/AIDS epidemic. *Vox*. December 1, 2016. www.vox.com/2015/12/1/9828348/ronald-reagan-hiv-aids. Accessed May 19, 2019.

35. Villarica H. 30 years of AIDS: 6,200 iconic posters, 100 countries, 1 collector. *Atlantic*. November 30, 2011; These posters show what AIDS meant in the 1980s. BuzzFeed.News. December 1, 2015. www.buzzfeed news.com/article/patrickstrudwick/these-1980s-aids-posters-show-the-desperate-fight-to-save-li. Accessed March 26, 2019; Schulman S. When protest movements became brands. *New York Times*. April 16, 2018.

36. Cohen MS, Chen YQ, McCauley M, et al. Prevention of HIV-1 infection with early antiretroviral therapy. *New England Journal of Medicine*. 2011;365(6):493–505; Treatment as prevention (TASP) for HIV. AVERT. www.avert.org/professionals/hiv-programming/prevention/treatment-as-prevention. Updated August 15, 2017. Accessed September 20, 2018.

37. Roser M, Ritchie H. HIV/AIDS. Our World in Data. November 2014. https://ourworldindata.org/hiv-aids. Updated April 2018. Accessed March 31, 2019.

38. When doctors refuse to treat AIDS. *New York Times*. August 3, 1987:A00016.

39. France D. *How to Survive a Plague: The Inside Story of How Citizens and Science Tamed AIDS*. Knopf; 2016.

40. Chang K. Q&A with David France. *Anthem*. September 20, 2012.

41. France D. *How to Survive a Plague: The Inside Story of How Citizens and Science Tamed AIDS*. Knopf; 2016.

42. Ryan's story. Ryan White 1971–1990. https://ryanwhite.com/Ryans_Story.html. Accessed March 27, 2019.

43. Ryan White CARE Act. Wikipedia. https://en.wikipedia.org/wiki/Ryan_White_CARE_Act. Accessed September 20, 2018.

44. The AIDS Memorial Quilt. The NAMES Project Foundation. www.aidsquilt.org/about/the-aids-memorial-quilt. Accessed September 2018; Jones C. How one man's idea for the AIDS quilt made the country pay attention. *Washington Post.* October 9, 2016; The AIDS Memorial Quilt. World Quilts: The American Story. http://worldquilts.quiltstudy.org/americanstory/engagement/NAMESquilt. Accessed March 5, 2019; Licostie N. The last one: the story of the AIDS Memorial Quilt. *POZ.* February 4, 2014. www.poz.com/article/nadine-licostie-25129-4218. Accessed March 5, 2019.

45. The AIDS Memorial Quilt. The NAMES Project Foundation. www.aidsquilt.org/about/the-aids-memorial-quilt. Accessed September 2018.

46. France D. *How to Survive a Plague: The Inside Story of How Citizens and Science Tamed AIDS.* Knopf; 2016.

47. The AIDS Memorial Quilt. The NAMES Project Foundation. www.aidsquilt.org/about/the-aids-memorial-quilt. Accessed September 2018.

48. Park A. The story behind the first AIDS drug. *Time.* March 19, 2017.

49. Ibid.

50. France D. How to survive a plague [film]. Sundance Selects. January 22, 2012. www.amazon.com/How-Survive-Plague-Peter-Staley/dp/B00A92MGLA; Peter Staley. Wikipedia. https://en.wikipedia.org/wiki/Peter_Staley. Accessed September 20, 2018.

51. Murphy T. Where are they now? ACT UP AIDS activists 25 years later. New York. June 25, 2013. http://nymag.com/intelligencer/2013/06/act-up-aids-activists-25-years-later.html. Accessed March 27, 2019.

52. Crimp D. Before Occupy: how AIDS activists seized control of the FDA in 1988. *Atlantic.* December 6, 2011.

53. France D. How to survive a plague [film]. Sundance Selects. January 22, 2012. www.amazon.com/How-Survive-Plague-Peter-Staley/dp/B00A92MGLA; Crimp D. Before Occupy: how AIDS activists seized control of the FDA in 1988. *Atlantic.* December 6, 2011.

54. Estimates of funding for various research, condition, and disease categories (RCDC). National Institutes of Health. 2018. https://report.nih.gov/categorical_spending.aspx. Updated May 18, 2018. Accessed March 31, 2019.

55. FDA approval of HIV medicines. US Department of Health and Human Services. AIDSinfo. https://aidsinfo.nih.gov/understanding-hiv-aids

/infographics/25/fda-approval-of-hiv-medicines. Updated September 20, 2018. Accessed December 3, 2018.

56. Land E. Why do some HIV drugs cost so much? Pharma, insurers, advocacy groups and consumers weigh in. *BETA Blog.* October 28, 2015. www.sfaf.org/collections/beta/why-are-hiv-meds-so-expensive-and-what -can-we-do-about-it. Accessed June 27, 2019.

57. Jesse Helms. Wikipedia. https://en.wikipedia.org/wiki/Jesse_Helms. Accessed September 21, 2018.

58. France D. *How to Survive a Plague: The Inside Story of How Citizens and Science Tamed AIDS.* Knopf; 2016.

59. In memory of Jesse Helms, and the condom on his house. July 8, 2008. POZ. www.poz.com/blog/in-memory-of-je. Accessed September 21, 2018.

60. Ibid.

61. Ibid.

62. Ibid.

63. Accelerating towards 90-90-90. UNAIDS. July 24, 2018. www.unaids .org/en/resources/presscentre/featurestories/2018/july/90-90-90-targets -workshop. Accessed September 21, 2018.

64. Ibid.

65. Schneider AP, Zainer CM, Kubat CK, Mullen NK, Windisch AK. The breast cancer epidemic: 10 facts. *Linacre Quarterly.* 2014;81(3):244– 277; King S. *Pink Ribbons, Inc.: Breast Cancer and the Politics of Philanthropy.* 1st ed. University of Minnesota Press; 2008; Rising global cancer epidemic. American Cancer Society. 2018. www.cancer.org/research/info graphics-gallery/rising-global-cancer-epidemic.html. Accessed March 31, 2019.

66. Lerner BH. Not so simple: the breast cancer stories of Betty Ford and Happy Rockefeller. *Huffington Post.* September 26, 2014.

67. Betty Ford. Wikipedia. https://en.wikipedia.org/wiki/Betty_Ford. Accessed September 21, 2018.

68. Lerner BH. Not so simple: the breast cancer stories of Betty Ford and Happy Rockefeller. *Huffington Post.* September 26, 2014.

69. Betty Ford. Wikipedia. https://en.wikipedia.org/wiki/Betty_Ford. Accessed September 21, 2018.

70. Osuch JR, Silk K, Price C, et al. A historical perspective on breast cancer activism in the United States: from education and support to partnership in scientific research. *Journal of Women's Health.* 2012;21(3):355–362.

71. Lorde A. *A Burst of Light.* Firebrand Books; 1988.

72. King S. *Pink Ribbons, Inc.: Breast Cancer and the Politics of Philanthropy*. 1st ed. University of Minnesota Press; 2008; Lorde A. *The Cancer Journals: Special Edition*. Aunt Lute Books; 2006.

73. Osuch JR, Silk K, Price C, et al. A historical perspective on breast cancer activism in the United States: from education and support to partnership in scientific research. *Journal of Women's Health*. 2012;21(3):355–362.

74. Komen Race for the Cure. Susan G. Komen Minnesota. 2018. www .komenminnesota.org/komen_race_for_the_cure_.html. Accessed March 5, 2019.

75. *Annual Report Fiscal Year 2017*. Susan G. Komen; 2018.

76. In memoriam: Charlotte Haley, creator of the first (peach) breast cancer ribbon. Breast Cancer Action. June 24, 2014. www.bcaction.org/2014/06 /24/in-memoriam-charlotte-haley-creator-of-the-first-peach-breast-cancer -ribbon. Accessed March 5, 2019.

77. King S. *Pink Ribbons, Inc.: Breast Cancer and the Politics of Philanthropy*. 1st ed. University of Minnesota Press; 2008; Quinn A. Before pink became synonymous with breast cancer, there was peach. Podcast: 8:21. *WHYY*. July 23,2015.https://whyy.org/segments/before-pink-became-synonymous-with -breast-cancer-there-was-peach. Accessed March 31, 2019.

78. King S. *Pink Ribbons, Inc.: Breast Cancer and the Politics of Philanthropy*. 1st ed. University of Minnesota Press; 2008; Quinn A. Before pink became synonymous with breast cancer, there was peach. Podcast: 8:21. *WHYY*. July 23, 2015. https://whyy.org/segments/before-pink-became -synonymous-with-breast-cancer-there-was-peach. Accessed March 31, 2019.

79. King S. Pink Ribbons Inc: breast cancer activism and the politics of philanthropy. *International Journal of Qualitative Studies in Education*. 2004;17(4):473–492.

80. Estimates of funding for various research, condition, and disease categories (RCDC). National Institutes of Health. 2018. https://report.nih.gov/ categorical_spending.aspx. Updated May 18, 2018. Accessed March 31, 2019.

81. *Breast Cancer Facts & Figures 2017–2018*. American Cancer Society; 2017. www.cancer.org/content/dam/cancer-org/research/cancer-facts-and -statistics/breast-cancer-facts-and-figures/breast-cancer-facts-and-figures -2017-2018.pdf; US Breast Cancer Statistics. Breastcancer.org. 2018. www .breastcancer.org/symptoms/understand_bc/statistics. Last updated February 13, 2019. Accessed March 31, 2019.

CHAPTER 4: BEFORE IT STARTS

1. Carson R, Darling L, Darling L. *Silent Spring*. Houghton Mifflin; 1962.

2. DDT—a brief history and status. US Environmental Protection Agency. www.epa.gov/ingredients-used-pesticide-products/ddt-brief-history-and-status. Updated August 11, 2017. Accessed March 25, 2019.

3. Public health statement for DDT, DDE, and DDD. Agency for Toxic Substances and Disease Registry. Toxic Substances Portal. September 2002. www.atsdr.cdc.gov/phs/phs.asp?id=79&tid=20. Accessed March 31, 2019.

4. Fleming L, Mann JB, Bean J, Briggle T, Sanchez-Ramos JR. Parkinson's disease and brain levels of organochlorine pesticides. *Annals of Neurology.* 1994;36(1):100–103; Weisskopf MG, Knekt P, O'Reilly EJ, et al. Persistent organochlorine pesticides in serum and risk of Parkinson disease. *Neurology.* 2010;74(13):1055–1061; Freire C, Koifman S. Pesticide exposure and Parkinson's disease: epidemiological evidence of association. *NeuroToxicology.* 2012;33(5):947–971.

5. Public health statement for DDT, DDE, and DDD. Agency for Toxic Substances and Disease Registry. Toxic Substances Portal. September 2002. www.atsdr.cdc.gov/phs/phs.asp?id=79&tid=20. Accessed March 31, 2019; Saeedi Saravi SS, Dehpour AR. Potential role of organochlorine pesticides in the pathogenesis of neurodevelopmental, neurodegenerative, and neurobehavioral disorders: a review. *Life Sciences.* 2016;145:255–264; Rossi M, Scarselli M, Fasciani I, Maggio R, Giorgi F. Dichlorodiphenyltrichloroethane (DDT) induced extracellular vesicle formation: a potential role in organochlorine increased risk of Parkinson's disease. *Acta Neurobiologiae Experimentalis.* 2017;77(2):113–117.

6. Gorell JM, Johnson CC, Rybicki BA, Peterson EL, Richardson RJ. The risk of Parkinson's disease with exposure to pesticides, farming, well water, and rural living. *Neurology.* 1998;50(5):1346–1350; Tüchsen F, Jensen AA. Agricultural work and the risk of Parkinson's disease in Denmark, 1981–1993. *Scandinavian Journal of Work, Environment & Health.* 2000;26(4):359–362; Elbaz A, Clavel J, Rathouz PJ, et al. Professional exposure to pesticides and Parkinson disease. *Annals of Neurology.* 2009;66(4):494–504.

7. Gorell JM, Johnson CC, Rybicki BA, Peterson EL, Richardson RJ. The risk of Parkinson's disease with exposure to pesticides, farming, well water, and rural living. *Neurology.* 1998;50(5):1346–1350.

8. Elbaz A, Clavel J, Rathouz PJ, et al. Professional exposure to pesticides and Parkinson disease. *Annals of Neurology.* 2009;66(4):494–504.

9. Kab S, Spinosi J, Chaperon L, et al. Agricultural activities and the incidence of Parkinson's disease in the general French population. *European Journal of Epidemiology.* 2017;32(3):203–216; Strickland D, Bertoni

JM. Parkinson's prevalence estimated by a state registry. *Movement Disorders*. 2004;19(3):318–323; Martino R, Candundo H, Lieshout PV, Shin S, Crispo JAG, Barakat-Haddad C. Onset and progression factors in Parkinson's disease: a systematic review. *NeuroToxicology*. 2017;61:132–141.

10. Ames RG, Howd RA, Doherty L. Community exposure to a paraquat drift. *Archives of Environmental Health: An International Journal*. 1993;48(1):47–52.

11. Martino R, Candundo H, Lieshout PV, Shin S, Crispo JAG, Barakat-Haddad C. Onset and progression factors in Parkinson's disease: a systematic review. *NeuroToxicology*. 2017;61:132–141; Gatto NM, Cockburn M, Bronstein J, Manthripragada AD, Ritz B. Well-water consumption and Parkinson's disease in rural California. *Environmental Health Perspectives*. 2009;117(12):1912–1918; James KA, Hall DA. Groundwater pesticide levels and the association with Parkinson disease. *International Journal of Toxicology*. 2015;34(3):266–273.

12. Gatto NM, Cockburn M, Bronstein J, Manthripragada AD, Ritz B. Well-water consumption and Parkinson's disease in rural California. *Environmental Health Perspectives*. 2009;117(12):1912–1918; Drinking water and pesticides. US Environmental Protection Agency. www.epa.gov /safepestcontrol/drinking-water-and-pesticides. Updated June 19, 2017. Accessed March 25, 2019.

13. Private drinking water wells. US Environmental Protection Agency. www.epa.gov/privatewells. Updated June 6, 2018. Accessed March 25, 2019; Private ground water wells. Centers for Disease Control and Prevention. www .cdc.gov/healthywater/drinking/private/wells/index.html. Last reviewed December 16, 2014. Accessed March 25, 2019; Water and pesticides. National Pesticide Information Center. http://npic.orst.edu/envir/water.html. Last updated January 13, 2016. Accessed March 25, 2019.

14. Strickland D, Bertoni JM. Parkinson's prevalence estimated by a state registry. *Movement Disorders*. 2004;19(3):318–323.

15. Barbeau A, Roy M, Bernier G, Campanella G, Paris S. Ecogenetics of Parkinson's disease: prevalence and environmental aspects in rural areas. *Canadian Journal of Neurological Sciences*. 1987;14(1):36–41.

16. Kab S, Spinosi J, Chaperon L, et al. Agricultural activities and the incidence of Parkinson's disease in the general French population. *European Journal of Epidemiology*. 2017;32(3):203–216.

17. Haspel T. The truth about organic produce and pesticides. *Daily Gazette*. May 23, 2018.

18. DDT (general fact sheet). National Pesticide Information Center; 1999. http://npic.orst.edu/factsheets/ddtgen.pdf.

19. Berry-Caban CS. DDT and Silent Spring: fifty years after. *Journal of Military and Veterans' Health*. 2011;19(4):19–24.

20. Bate R. *The Rise, Fall, Rise, and Imminent Fall of DDT*. American Enterprise Institute for Public Policy Research; November 5, 2007.

21. Whorton JC. *Before Silent Spring: Pesticides and Public Health in Pre-DDT America*. Princeton University Press; 2015.

22. DDT. Wikipedia. https://en.wikipedia.org/wiki/DDT. Accessed October 15, 2018; Rogan WJ, Chen A. Health risks and benefits of bis(4-chlorophenyl)-1,1,1-trichloroethane (DDT). *The Lancet*. 2005;366(9487):763–773; Dr. Paul Müller. *Nature*. 1965;208(5015):1043–1044; Paul Hermann Müller. Wikipedia. https://en.wikipedia.org/wiki/Paul_Hermann_M%C3%BCller. Accessed March 25, 2019.

23. Longnecker MP, Rogan WJ, Lucier G. The human health effects of DDT (dichlorodiphenyltrichloroethane) and PCBS (polychlorinated biphenyls) and an overview of organochlorines in public health. *Annual Review of Public Health*. 1997;18(1):211–244.

24. Snedeker SM. Pesticides and breast cancer risk: a review of DDT, DDE, and dieldrin. *Environmental Health Perspectives*. 2001;109(Suppl 1):35–47.

25. Longnecker MP, Rogan WJ, Lucier G. The human health effects of DDT (dichlorodiphenyltrichloroethane) and PCBS (polychlorinated biphenyls) and an overview of organochlorines in public health. *Annual Review of Public Health*. 1997;18(1):211–244.

26. Dichlorodiphenyltrichloroethane (DDT). Centers for Disease Control and Prevention; 2009. www.cdc.gov/biomonitoring/pdf/ddt_factsheet.pdf.

27. Pesticide information profile: DDT (dichlorodiphenyltrichloroethane). Extension Toxciology Network. http://pmep.cce.cornell.edu/profiles/extoxnet/carbaryl-dicrotophos/ddt-ext.html. Accessed March 25, 2019.

28. Stellman JM, Stellman SD, Christian R, Weber T, Tomasallo C. The extent and patterns of usage of Agent Orange and other herbicides in Vietnam. *Nature*. 2003;422:681–687.

29. Ibid.

30. Stone R. Agent Orange's bitter harvest. *Science*. 2007;315(5809):176–179.

31. Yi S-W, Ohrr H, Hong J-S, Yi J-J. Agent Orange exposure and prevalence of self-reported diseases in Korean Vietnam veterans. *Journal of Preventive Medicine and Public Health*. 2013;46(5):213–225; Medicine Io. *Veterans and Agent Orange: Update 2010*. National Academies Press; 2012.

32. Parkinson's disease and Agent Orange. US Department of Veterans Affairs: Public Health. www.publichealth.va.gov/exposures/agentorange/conditions/parkinsonsdisease.asp. Accessed March 25, 2019.

33. Hawaii recalls pesticide-laced milk from stores and schools. *New York Times*. March 20, 1982.

34. Contaminated milk problem in Hawaii nears end. *New York Times*. May 23, 1982; Smith RJ. Hawaiian milk contamination creates alarm. *Science*. 1982;217(4555):137–140.

35. Smith RJ. Hawaiian milk contamination creates alarm. *Science*. 1982;217(4555):137–140.

36. Ibid.

37. Hawaii recalls pesticide-laced milk from stores and schools. *New York Times*. March 20, 1982.

38. Ibid.; Contaminated milk problem in Hawaii nears end. *New York Times*. May 23, 1982.

39. Smith RJ. Hawaiian milk contamination creates alarm. *Science*. 1982;217(4555):137–140.

40. Ibid.

41. Ibid.

42. Abbott RD, Ross GW, Petrovitch H, et al. Midlife milk consumption and substantia nigra neuron density at death. *Neurology*. 2016;86(6):512–519.

43. Ibid.

44. Ibid.

45. Park A. Drinking milk is linked to Parkinson's disease: study. *Time*. December 9, 2015.

46. Aldrin/dieldrin. Agency for Toxic Substances and Disease Registry. Toxic Substances Portal. www.atsdr.cdc.gov/substances/toxsubstance.asp ?toxid=56. Last updated March 3, 2011. Accessed March 25, 2019; Aldrin and Dieldrin—ToxFAQs. Agency for Toxic Substances and Disease Registry; 2002. www.atsdr.cdc.gov/toxfaqs/tfacts1.pdf.

47. Fleming L, Mann JB, Bean J, Briggle T, Sanchez-Ramos JR. Parkinson's disease and brain levels of organochlorine pesticides. *Annals of Neurology*. 1994;36(1):100–103; Richardson JR, Shalat SL, Buckley B, et al. Elevated serum pesticide levels and risk of Parkinson disease. *Archives of Neurology*. 2009;66(7):870–875; Corrigan FM, Wienburg CL, Shore RF, Daniel SE, Mann D. Organochlorine insecticides in substantia nigra in Parkinson's disease. *Journal of Toxicology and Environmental Health, Part A*. 2000;59(4):229–234.

48. Sanchez-Ramos J, Facca A, Basit A, Song S. Toxicity of dieldrin for dopaminergic neurons in mesencephalic cultures. *Experimental Neurology*. 1998;150(2):263–271; Richardson JR, Caudle WM, Wang M, Dean ED, Pennell KD, Miller GW. Developmental exposure to the pesticide dieldrin

alters the dopamine system and increases neurotoxicity in an animal model of Parkinson's disease. *FASEB Journal*. 2006;20(10):1695–1697.

49. Wong MH, Leung AOW, Chan JKY, Choi MPK. A review on the usage of POP pesticides in China, with emphasis on DDT loadings in human milk. *Chemosphere*. 2005;60(6):740–752.

50. Ibid.

51. Ibid.; Sifuentes Dos Santos J, Schwanz TG, Coelho AN, et al. Estimated daily intake of organochlorine pesticides from dairy products in Brazil. *Food Control*. 2015;53:23–28; Kampire E, Kiremire BT, Nyanzi SA, Kishimba M. Organochlorine pesticide in fresh and pasteurized cow's milk from Kampala markets. *Chemosphere*. 2011;84(7):923–927; Gebremichael S, Birhanu T, Tessema DA. Analysis of organochlorine pesticide residues in human and cow's milk in the towns of Asendabo, Serbo and Jimma in south-western Ethiopia. *Chemosphere*. 2013;90(5):1652–1657.

52. Vall O, Gomez-Culebras M, Puig C, et al. Prenatal and postnatal exposure to DDT by breast milk analysis in Canary Islands. *PLOS One*. 2014;9(1):e83831–e83831; Kanthasamy AG, Kitazawa M, Kanthasamy A, Anantharam V. Dieldrin-induced neurotoxicity: relevance to Parkinson's disease pathogenesis. *NeuroToxicology*. 2005;26(4):701–719; Chen M-W, Santos HM, Que DE, et al. Association between organochlorine pesticide levels in breast milk and their effects on female reproduction in a Taiwanese population. *International Journal of Environmental Research and Public Health*. 2018;15(5):931.

53. Jorgenson JL. Aldrin and dieldrin: a review of research on their production, environmental deposition and fate, bioaccumulation, toxicology, and epidemiology in the United States. *Environmental Health Perspectives*. 2001;109(Suppl 1):113–139.

54. Saeedi Saravi SS, Dehpour AR. Potential role of organochlorine pesticides in the pathogenesis of neurodevelopmental, neurodegenerative, and neurobehavioral disorders: a review. *Life Sciences*. 2016;145:255–264; Pesticide information profile: DDT (dichlorodiphenyltrichloroethane). Extension Toxiology Network. http://pmep.cce.cornell.edu/profiles/extoxnet /carbaryl-dicrotophos/ddt-ext.html. Accessed March 25, 2019; Gannon N, Link RP, Decker GC. Pesticide residues in milk, insecticide residues in the milk of dairy cows fed insecticides in their daily ration. *Journal of Agricultural and Food Chemistry*. 1959;7(12):829–832.

55. Liu J, Morrow AL, Devaud LL, Grayson DR, Lauder JM. Regulation of GABAA receptor subunit mRNA expression by the pesticide dieldrin in embryonic brainstem cultures: a quantitative, competitive reverse

transcription-polymerase chain reaction study. *Journal of Neuroscience Research.* 1997;49(5):645–653.

56. Saeedi Saravi SS, Dehpour AR. Potential role of organochlorine pesticides in the pathogenesis of neurodevelopmental, neurodegenerative, and neurobehavioral disorders: a review. *Life Sciences.* 2016;145:255–264.

57. Tanner CM, Kamel F, Ross GW, et al. Rotenone, paraquat, and Parkinson's disease. *Environmental Health Perspectives.* 2011;119(6):866–872; Dinis-Oliveira RJ, Remião F, Carmo H, et al. Paraquat exposure as an etiological factor of Parkinson's disease. *NeuroToxicology.* 2006;27(6):1110–1122; Thiruchelvam M, McCormack A, Richfield EK, et al. Age-related irreversible progressive nigrostriatal dopaminergic neurotoxicity in the paraquat and maneb model of the Parkinson's disease phenotype. *European Journal of Neuroscience.* 2003;18(3):589–600; Wang A, Costello S, Cockburn M, Zhang X, Bronstein J, Ritz B. Parkinson's disease risk from ambient exposure to pesticides. *European Journal of Epidemiology.* 2011;26(7):547–555.

58. Hakim D. This pesticide is prohibited in Britain. Why is it still being exported? *New York Times.* December 20, 2016.

59. Paraquat benefits. Paraquat Information Center. https://paraquat .com/en/benefits. Accessed October 15, 2018.

60. Pesticide National Synthesis Project: estimated annual agricultural pesticide use. US Geological Survey. https://water.usgs.gov/nawqa/pnsp/usage /maps/show_map.php?year=2015&map=PARAQUAT&hilo=H. Updated September 11, 2018. Accessed March 25, 2019.

61. Ibid.

62. Ibid.

63. Ibid.

64. Costello S, Cockburn M, Bronstein J, Zhang X, Ritz B. Parkinson's disease and residential exposure to maneb and paraquat from agricultural applications in the Central Valley of California. *American Journal of Epidemiology.* 2009;169(8):919–926.

65. Tanner CM, Kamel F, Ross GW, et al. Rotenone, paraquat, and Parkinson's disease. *Environmental Health Perspectives.* 2011;119(6):866–872; Goldman SM, Kamel F, Ross GW, et al. Genetic modification of the association of paraquat and Parkinson's disease. *Movement Disorders.* 2012;27(13):1652–1658.

66. Hakim D. This pesticide is prohibited in Britain. Why is it still being exported? *New York Times.* December 20, 2016.

67. Betarbet R, Sherer TB, Greenamyre JT. Animal models of Parkinson's disease. *BioEssays.* 2002;24(4):308–318.

68. Brooks AI, Chadwick CA, Gelbard HA, Cory-Slechta DA, Federoff HJ. Paraquat elicited neurobehavioral syndrome caused by dopaminergic neuron loss. *Brain Research*. 1999;823(1):1–10.

69. Watts M. *Paraquat*. Pesticide Action Network Asia & the Pacific. February 2011. www.panna.org/sites/default/files/Paraquat%20monograph %20final%202011-1.pdf. Accessed June 28, 2019.

70. Joyce M. Ocular damage caused by paraquat. *British Journal of Ophthalmology*. 1969;53(10):688–690.

71. Pesticide information profile: paraquat. Extension Toxicology Network. September 1993. http://pmep.cce.cornell.edu/profiles/extoxnet/metiram -propoxur/paraquat-ext.html. Accessed March 25, 2019.

72. Cha ES, Lee WJ, Chang S-S, Gunnell D, Eddleston M, Khang Y-H. Impact of paraquat regulation on suicide in South Korea. *International Journal of Epidemiology*. 2015;45(2):470–479; Lin C, Yen T-H, Juang Y-Y, Lee C-P, Lee S-H. Distinct psychopathology of patients who attempted suicide with rodenticide in Taiwan: a comparative study with patients of suicide with paraquat. *Psychology Research and Behavior Management*. 2018;11:323–328.

73. Hakim D. This pesticide is prohibited in Britain. Why is it still being exported? *New York Times*. December 20, 2016; Mercola J. Paraquat—banned in EU while US increasing use of this toxic killer. Mercola: Take Control of Your Health. January 3, 2017. https://articles.mercola.com/sites/articles/archive /2017/01/03/paraquat-banned-in-32-countries.aspx. Accessed May 20, 2019.

74. Hakim D. This pesticide is prohibited in Britain. Why is it still being exported? *New York Times*. December 20, 2016.

75. Ibid.

76. Watts M. *Paraquat*. Pesticide Action Network Asia & the Pacific. February 2011. www.panna.org/sites/default/files/Paraquat%20monograph %20final%202011-1.pdf. Accessed June 28, 2019.

77. Hakim D. This pesticide is prohibited in Britain. Why is it still being exported? *New York Times*. December 20, 2016.

78. FTC requires China National Chemical Corporation and Syngenta AG to divest U.S. assets as a condition of merger [press release]. Federal Trade Commission, April 4, 2017; McConnell W. ChemChina gets FTC OK to snap up Syngenta. *TheStreet*. April 4, 2017. www.thestreet.com/story /14073559/1/chemchina-gets-ftc-ok-to-snap-up-syngenta.html. Accessed September 20, 2018.

79. Concern over paraquat. "Re: Docket EPA-HQ-OPP-2011-0855 (paraquat dichloride registration review)." Michael J. Fox Foundation. July

24, 2017. https://files.michaeljfox.org/Paraquat_letter_FINAL.pdf; personal communication.

80. Paraquat dichloride: reregistration eligibility decision (RED) fact sheet. US Environmental Protection Agency. August 1997. https://archive .epa.gov/pesticides/reregistration/web/pdf/0262fact.pdf.

81. Paraquat Dichloride: One Sip Can Kill. US Environmental Protection Agency. Pesticide Worker Safety. www.epa.gov/pesticide-worker-safety /paraquat-dichloride-one-sip-can-kill. Last updated May 8, 2019. Accessed March 25, 2019.

82. Ibid.

83. Ibid.

84. Ibid.

85. The Michael J. Fox Foundation submits 107,000 community signatures urging the EPA to ban herbicide linked to Parkinson's [press release]. The Michael J. Fox Foundation for Parkinson's Research, February 6, 2019.

86. Ibid.

87. Pesticide National Synthesis Project: estimated annual agricultural pesticide use. US Geological Survey. https://water.usgs.gov/nawqa/pnsp /usage/maps/show_map.php?year=2015&map=PARAQUAT&hilo=H. Updated September 11, 2018. Accessed March 25, 2019.

88. Paraquat dichloride. US Environmental Protection Agency. www.epa .gov/ingredients-used-pesticide-products/paraquat-dichloride. Updated May 15, 2019. Accessed May 19, 2019.

89. Pesticide National Synthesis Project: estimated annual agricultural pesticide use. US Geological Survey. https://water.usgs.gov/nawqa/pnsp /usage/maps/show_map.php?year=2015&map=PARAQUAT&hilo=H. Updated September 11, 2018. Accessed March 25, 2019.

90. Furlong M, Tanner CM, Goldman SM, et al. Protective glove use and hygiene habits modify the associations of specific pesticides with Parkinson's disease. *Environment International.* 2015;75:144–150.

91. Doll R, Hill AB. Lung cancer and other causes of death in relation to smoking; a second report on the mortality of British doctors. *BMJ.* 1956;2(5001):1071–1081; Doll R, Hill AB. The mortality of doctors in relation to their smoking habits: a preliminary report. *BMJ.* 1954;1(4877):1451–1455; Hill AB. The environment and disease: association or causation? *Proceedings of the Royal Society of Medicine.* 1965;58(5):295–300.

92. Hill AB. The environment and disease: association or causation? *Proceedings of the Royal Society of Medicine.* 1965;58(5):295–300; Fedak KM, Bernal A, Capshaw ZA, Gross S. Applying the Bradford Hill criteria in the

21st century: how data integration has changed causal inference in molecular epidemiology. *Emerging Themes in Epidemiology.* 2015;12:14.

93. Bates C, Rowell A. *Tobacco Explained: The Truth About the Tobacco Industry.* Action on Smoking and Health; 1998.

94. Hakim D, Lipton E. Pesticide studies won E.P.A.'s trust until Trump's team scorned "secret science." *New York Times.* August 24, 2018.

95. Betarbet R, Sherer TB, MacKenzie G, Garcia-Osuna M, Panov AV, Greenamyre JT. Chronic systemic pesticide exposure reproduces features of Parkinson's disease. *Nature Neuroscience.* 2000;3(12):1301–1306; Sparling AS, Martin DW, Posey LB. An evaluation of the proposed worker protection standard with respect to pesticide exposure and Parkinson's disease. *International Journal of Environmental Research and Public Health.* 2017;14(6):640; Van Maele-Fabry G, Hoet P, Vilain F, Lison D. Occupational exposure to pesticides and Parkinson's disease: a systematic review and meta-analysis of cohort studies. *Environment International.* 2012;46:30–43; Gunnarsson L-G, Bodin L. Parkinson's disease and occupational exposures: a systematic literature review and meta-analyses. *Scandinavian Journal of Work, Environment & Health.* 2017(3):197–209; Johnson ME, Bobrovskaya L. An update on the rotenone models of Parkinson's disease: their ability to reproduce the features of clinical disease and model gene-environment interactions. *NeuroToxicology.* 2015;46:101–116; Petrovitch H, Ross G, Abbott RD, et al. Plantation work and risk of Parkinson disease in a population-based longitudinal study. *Archives of Neurology.* 2002;59(11):1787–1792.

96. Breckenridge CB, Berry C, Chang ET, Sielken RL, Jr., Mandel JS. Association between Parkinson's disease and cigarette smoking, rural living, well-water consumption, farming and pesticide use: systematic review and meta-analysis. *PLOS One.* 2016;11(4):e0151841–e0151841.

97. Ibid.

98. Carson R, Darling L, Darling L. *Silent Spring.* Houghton Mifflin; 1962.

99. Ibid.

100. Ibid.

101. Aktar MW, Sengupta D, Chowdhury A. Impact of pesticides use in agriculture: their benefits and hazards. *Interdisciplinary Toxicology.* 2009;2(1):1–12; Cooper J, Dobson H. The benefits of pesticides to mankind and the environment. *Crop Protection.* 2007;26(9):1337–1348; Shelton JF, Hertz-Picciotto I, Pessah IN. Tipping the balance of autism risk: potential mechanisms linking pesticides and autism. *Environmental Health Perspectives.* 2012;120(7):944–951; Shelton JF, Geraghty EM, Tancredi DJ, et al. Neurodevelopmental disorders and prenatal residential proximity to agri-

cultural pesticides: the CHARGE Study. *Environmental Health Perspectives.* 2014;122(10):1103–1109; Holzman DC. Pesticides and autism spectrum disorders: new findings from the CHARGE Study. *Environmental Health Perspectives.* 2014;122(10):A280–A280; Jee S-H, Kuo H-W, Su WPD, Chang C-H, Sun C-C, Wang J-D. Photodamage and skin cancer among paraquat workers. *International Journal of Dermatology.* 1995;34(7):466–469; Pesticide information profile: paraquat. Extension Toxicology Network. September 1993. http://pmep.cce.cornell.edu/profiles/extoxnet/metiram-propoxur/paraquat-ext.html. Accessed March 25, 2019.

102. Gatto NM, Cockburn M, Bronstein J, Manthripragada AD, Ritz B. Well-water consumption and Parkinson's disease in rural California. *Environmental Health Perspectives.* 2009;117(12):1912–1918; Wang A, Costello S, Cockburn M, Zhang X, Bronstein J, Ritz B. Parkinson's disease risk from ambient exposure to pesticides. *European Journal of Epidemiology.* 2011;26(7):547–555.

CHAPTER 5: CLEANING UP

1. Hardmon T, Libert R. Semper fi: always faithful [film]. April 21, 2011.

2. Goldman SM, Quinlan PJ, Ross GW, et al. Solvent exposures and Parkinson disease risk in twins. *Annals of Neurology.* 2012;71(6):776–784.

3. IARC Working Group on the Evaluation of Carcinogenic Risk to Humans. Trichloroethylene, tetrachloroethylene, and some other chlorinated Agents. IARC Monographs on the Evaluation of Carcinogenic Risks to Humans 106. Lyon: International Agency for Research on Cancer; 2014. Trichloroethylene. IDC Technologies. www.idc-online.com/technical_references/pdfs/chemical_engineering/Trichloroethylene.pdf. Accessed March 29, 2019.

4. ToxFAQs for trichloroethylene (TCE). Agency for Toxic Substances & Disease Registry. Toxic Substances Portal. www.atsdr.cdc.gov/toxfaqs/tf.asp?id=172&tid=30. Updated November 4, 2016. Accessed March 14, 2019; Trichloroethylene. Wikipedia. https://en.wikipedia.org/wiki/Trichloroethylene. Accessed March 14, 2019; Campbell-Dollaghan K. The secret history of Silicon Valley and the toxic remnants of the first computers. Gizmodo. November 20, 2015. https://gizmodo.com/the-secret-history-of-silicon-valley-and-the-toxic-remn-1743622225. Accessed March 14, 2019.

5. IARC Working Group on the Evaluation of Carcinogenic Risk to Humans. Trichloroethylene, tetrachloroethylene, and some other chlorinated agents. *IARC Monographs on the Evaluation of Carcinogenic Risks to Humans.* 2014;106:1–512.

6. Guehl D, Bezard E, Dovero S, Boraud T, Bioulac B, Gross C. Trichloro-ethylene and parkinsonism: a human and experimental observation. *European Journal of Neurology*. 1999;6(5):609–611.

7. TCE overview. Agency for Toxic Substances and Disease Registry. Trichloroethylene (TCE). www.atsdr.cdc.gov/tox-tool/trichloroethylene/01/tce_overview.html. Accessed March 29, 2019.

8. Risk management for trichloroethylene (TCE). US Environmental Protection Agency. www.epa.gov/assessing-and-managing-chemicals-under-tsca/risk-management-trichloroethylene-tce. Updated December 14, 2017. Accessed November 16, 2018.

9. James WR. Fatal addiction to trichloroethylene. *British Journal of Industrial Medicine*. 1963;20(1):47–49.

10. McCord CP. Toxicity of trichloroethylene. *Journal of the American Medical Association*. 1932;99(5):409.

11. Ibid.

12. Gash DM, Rutland K, Hudson NL, et al. Trichloroethylene: parkinsonism and complex 1 mitochondrial neurotoxicity. *Annals of Neurology*. 2008;63(2):184–192.

13. Ibid.

14. Ibid.

15. Ibid.

16. Ibid.

17. Barringer F. Exposed to solvent, worker faces hurdles. *New York Times*. January 24, 2009.

18. Barringer F. E.P.A. charts risks of a ubiquitous chemical. *New York Times*. September 30, 2011.

19. Goldman SM, Quinlan PJ, Ross GW, et al. Solvent exposures and Parkinson disease risk in twins. *Annals of Neurology*. 2012;71(6):776–784.

20. Ibid.

21. Freshwater LL. What happened at Camp Lejeune. In *Pacific Standard*. Social Justice Foundation; 2018; Marine Corps Base Camp Lejeune. Marines: The Official Website of the United States Marine Corps. www.lejeune.marines.mil/About. Accessed March 29, 2019.

22. Freshwater LL. What happened at Camp Lejeune. In *Pacific Standard*. Social Justice Foundation; 2018.

23. Goldman SM, Quinlan PJ, Ross GW, et al. Solvent exposures and Parkinson disease risk in twins. *Annals of Neurology*. 2012;71(6):776–784.

24. Nazaryan A. Camp Lejeune and the U.S. military's polluted legacy. *Newsweek*. July 16, 2014.

25. ABC One Hour Cleaners Jacksonville, NC: Cleanup Activities. US Environmental Protection Agency. Superfund Site. https://cumulis.epa .gov/supercpad/SiteProfiles/index.cfm?fuseaction=second.cleanup&id =0402718. Updated October 23, 2018. Accessed March 29, 2019.

26. Council NR. *Contaminated Water Supplies at Camp Lejeune: Assessing Potential Health Effects*. National Academies Press; 2009.

27. Goldman SM, Quinlan PJ, Ross GW, et al. Solvent exposures and Parkinson disease risk in twins. *Annals of Neurology*. 2012;71(6):776 784; Council NR. *Contaminated Water Supplies at Camp Lejeune: Assessing Potential Health Effects*. National Academies Press; 2009.

28. Freshwater LL. What happened at Camp Lejeune. In *Pacific Standard*. Social Justice Foundation; 2018; Camp Lejeune: Contamination and Compensation, Looking Back, Moving Forward. US Government Publishing Office; 2010. www.govinfo.gov/content/pkg/CHRG-111hhrg58485/html /CHRG-111hhrg58485.htm.

29. Peeples L. Camp Lejeune water contamination cover-up hinted in navy letter. Huffington Post. January 13, 2012. www.huffpost.com/entry/camp -lejeune-water-contamination-navy-letter_n_1203465. Updated December 6, 2017. Accessed March 29, 2019.

30. Smithberger M. Watchdog: significant concerns regarding drinking water safety at navy bases overseas. Project on Government Oversight. August 9, 2017. www.pogo.org/investigation/2017/08/watchdog-significant -concerns-regarding-drinking-water-safety-at-navy-bases-overseas. Accessed March 29, 2019; Wagner A, Maurer K. A legacy of pain for Camp Lejeune water contamination victims. GateHouse Media. Star News Online. March 25, 2017. www.starnewsonline.com/news/20170325/legacy-of-pain-for-camp -lejeune-water-contamination-victims. Accessed March 29, 2019.

31. Camp Lejeune: Contamination and Compensation, Looking Back, Moving Forward. US Government Publishing Office; 2010. www.govinfo .gov/content/pkg/CHRG-111hhrg58485/html/CHRG-111hhrg58485.htm.

32. Ibid.

33. Ibid.

34. Freshwater LL. What happened at Camp Lejeune. In *Pacific Standard*. Social Justice Foundation; 2018; Camp Lejeune drinking water U.S. Marine Corps Base Camp Lejeune, North Carolina. Agency for Toxic Substances and Disease Registry. December 20, 2017. www.atsdr.cdc.gov/HAC /pha/MarineCorpsBaseCampLejeune/Camp_Lejeune_Drinking_Water _PHA(final)_%201-20-2017_508.pdf; Pipe A, Freshwater L. *Poisoned by My Government: How the Military Let Families Drink Contaminated Water for*

Decades and Didn't Learn from It. Craig Newmark Graduate School of Journalism, City University of New York; 2017.

35. Freshwater LL. What happened at Camp Lejeune. In *Pacific Standard*. Social Justice Foundation; 2018.

36. Ibid.; Freshwater LL. Medium. https://medium.com/@LFreshwater. Accessed March 29, 2019.

37. Freshwater LL. What happened at Camp Lejeune. In *Pacific Standard*. Social Justice Foundation; 2018.

38. Schrade B. Years later, marine families bear scars of poisoning at Camp Lejeune. *Atlanta Journal-Constitution*. February 23, 2018.

39. Kime P. The toxic homefront: as marine families fall ill, some are accusing the corps of negligence. Task & Purpose. April 19, 2017. https://taskandpurpose.com/the-toxic-homefront-as-marine-families-fall-ill-some-are-accusing-the-corps-of-negligence. Accessed March 29, 2019.

40. Affairs DoV. Diseases associated with exposure to contaminants in the water supply at Camp Lejeune. *Federal Register*. 2017;82(9).

41. Ibid.

42. Fisher J. Poison valley. Salon. 2001. www.salon.com/2001/07/30/almaden1. Updated July 31, 2001. Accessed March 29, 2019.

43. Campbell-Dollaghan K. The secret history of Silicon Valley and the toxic remnants of the first computers. Gizmodo. November 20, 2015. https://gizmodo.com/the-secret-history-of-silicon-valley-and-the-toxic-remn-1743622225. Accessed March 14, 2019; Chepesiuk R. Where the chips fall: environmental health in the semiconductor industry. *Environmental Health Perspectives*. 1999;107(9):452–457.

44. Superfund sites in reuse in California. US Environmental Protection Agency. Superfund Redevelopment Initiative. www.epa.gov/superfund-redevelopment-initiative/superfund-sites-reuse-california. Updated March 7, 2019. Accessed March 29, 2019; Fairchild Semiconductor Corp. (Mountain View Plant) Mountain View, CA: cleanup activities. US Environmental Protection Agency. 2018. https://cumulis.epa.gov/supercpad/SiteProfiles/index.cfm?fuseaction=second.Cleanup&id=0901680. Accessed March 31, 2019.

45. What is Superfund? US Environmental Protection Agency. www.epa.gov/superfund/what-superfund. Updated November 30, 2018. Accessed March 28, 2019.

46. Ibid.; Nazaryan A. The US Department of Defense is one of the world's biggest polluters. *Newsweek*. July 17, 2014.

47. Superfund sites in reuse in California. US Environmental Protection Agency. Superfund Redevelopment Initiative. www.epa.gov/superfund

-redevelopment-initiative/superfund-sites-reuse-california. Updated March 7, 2019. Accessed March 29, 2019.

48. Fairchild Semiconductor Corp. (Mountain View Plant) Mountain View, CA: cleanup activities. US Environmental Protection Agency. 2018. https://cumulis.epa.gov/supercpad/SiteProfiles/index.cfm?fuseaction =second.Cleanup&id=0901680. Accessed March 31, 2019.

49. Horton J. I live next to Google—and on top of a toxic site. Don't let polluters be evil. *Guardian.* March 19, 2014.

50. Ibid.

51. Ibid.

52. DeBolt D. TCE causes cancer, other health problems, EPA says. *Mountain View Voice.* October 7, 2011; Hasan F. Residents near old semiconductor plant worry about toxic exposure. *Mountain View Voice.* November 22, 2002.

53. Risk management for trichloroethylene (TCE). US Environmental Protection Agency. www.epa.gov/assessing-and-managing-chemicals-under -tsca/risk-management-trichloroethylene-tce. Updated December 14, 2017. Accessed November 16, 2018; ToxFAQs for trichloroethylene (TCE). Agency for Toxic Substances & Disease Registry. Toxic Substances Portal. www.atsdr.cdc.gov/toxfaqs/tf.asp?id=172&tid=30. Updated November 4, 2016. Accessed March 14, 2019; DeBolt D. TCE causes cancer, other health problems, EPA says. *Mountain View Voice.* October 7, 2011.

54. Goldman SM, Quinlan PJ, Ross GW, et al. Solvent exposures and Parkinson disease risk in twins. *Annals of Neurology.* 2012;71(6):776–784; Bowdler N. Study links Parkinson's disease to industrial solvent. *BBC News.* November 14, 2011.

55. Superfund: National Priorities List. US Environmental Protection Agency. www.epa.gov/superfund/superfund-national-priorities-list-npl. Updated June 4, 2018. Accessed January 9, 2019.

56. Trichloroehtylene (TCE). Algona/Auburn Public Awareness Coalition. www.wa-apac.org/questions-from-the-community-factsheets/trichloroethylene -tce/. Accessed August 10, 2018.

57. Record of decision: Modock Road Springs/DLS Sand and Gravel, Inc. Site, Town of Victor, Ontario County, New York, Site Number 8-35-013. New York State Department of Environmental Conservation. January 2010. www.dec.ny.gov/docs/remediation_hudson_pdf/835013rod.pdf. Accessed March 29, 2019.

58. Ibid.; Modock Springs. Town of Victor, New York. www.victorny.org /260/Modock-Springs. Accessed March 29, 2019.

59. Youngman J, Maslanik M. Cancer study of Victor's "plume" inconclusive. *Daily Messenger*. October 16, 2009.

60. Barringer F. Exposed to solvent, worker faces hurdles. *New York Times*. January 24, 2009; Trichloroethylene. Wikipedia. https://en.wikipedia.org /wiki/Trichloroethylene. Updated May 28, 2018. Accessed March 31, 2019; C2 chlorinated solvents. IHS Markit. October 2017. https://ihsmarkit.com /products/c2-chlorinated-chemical-economics-handbook.html. Accessed March 26, 2019.

61. Risk management for trichloroethylene (TCE). US Environmental Protection Agency. www.epa.gov/assessing-and-managing-chemicals-under -tsca/risk-management-trichloroethylene-tce. Updated December 14, 2017. Accessed November 16, 2018; Integrated Risk Information System: trichloroethylene. US Environmental Protection Agency. https://cfpub.epa.gov/ncea /iris2/chemicalLanding.cfm?substance_nmbr=199. Updated July 28, 2017. Accessed March 28, 2019; Public health statement for trichloroethylene. Agency for Toxic Substances & Disease Registry. Toxic Substances Portal. www.atsdr.cdc.gov/phs/phs.asp?id=171&tid=30. Accessed March 29, 2019.

62. Risk management for trichloroethylene (TCE). US Environmental Protection Agency. www.epa.gov/assessing-and-managing-chemicals-under-tsca /risk-management-trichloroethylene-tce. Updated December 14, 2017. Accessed November 16, 2018.

63. Franklin K. Industry urges EPA to put the brakes on TCE ban. Chemical Watch. 2017. https://chemicalwatch.com/55210/industry-urges-epa-to -put-the-brakes-on-tce-ban. Updated 2018. Accessed March 31, 2019.

64. Ibid.

65. Ibid.

66. Kaplan S. In reversal, chemicals are cleared for use. *New York Times*. December 20, 2017:A16.

67. Ebbs S. Senators, families urge EPA to take stronger action on toxic chemicals. ABCNews. August 1, 2018. https://abcnews.go.com/Politics /senators-families-urge-epa-stronger-action-toxic-chemicals/story?id =56974123. Accessed December 6, 2018.

68. Hill AB. The environment and disease: association or causation? *Proceedings of the Royal Society of Medicine*. 1965;58(5):295–300.

69. GBD 2016 Parkinson's Disease Collaborators. Global, regional, and national burden of Parkinson's disease, 1990–2016: a systematic analysis for the Global Burden of Disease Study 2016. *Lancet Neurology*. 2018;17(11):939–953.

70. Hofman A, Stricker BH, Ikram MA, Koudstaal PJ, Darweesh SKL. Trends in the incidence of Parkinson disease in the general population: the Rotterdam Study. *American Journal of Epidemiology.* 2016;183(11):1018–1026.

71. Wesseling C, De Joode BVW, Ruepert C, et al. Paraquat in developing countries. *International Journal of Occupational and Environmental Health.* 2001;7(4):275–286.

72. Greve PA, Van Zoonen P. Organochlorine pesticides and PCBs in tissues from Dutch citizens (1968–1986). *International Journal of Environmental Analytical Chemistry.* 1990;38(2):265–277.

73. *Report on Carcinogens Background Document for Trichloroethylene.* National Toxicology Program; December 13–14, 2000.

74. Ritz B, Lee P-C, Hansen J, et al. Traffic-related air pollution and Parkinson's disease in Denmark: a case-control study. *Environmental Health Perspectives.* 2016;124(3):351–356; Finkelstein MM, Jerrett M. A study of the relationships between Parkinson's disease and markers of traffic-derived and environmental manganese air pollution in two Canadian cities. *Environmental Research.* 2007;104(3):420–432; Block ML, Calderón-Garcidueñas L. Air pollution: mechanisms of neuroinflammation and CNS disease. *Trends in Neurosciences.* 2009;32(9):506–516; Netherlands—air pollution country fact sheet 2018. European Environment Agency. www.eea.europa .eu/themes/air/country-fact-sheets/netherlands. Accessed March 29, 2019.

75. Netherlands—air pollution country fact sheet 2018. European Environment Agency. www.eea.europa.eu/themes/air/country-fact-sheets /netherlands. Accessed March 29, 2019.

CHAPTER 6: PROTECTING OURSELVES

1. Almendrala A. Muhammad Ali's greatest fight was against Parkinson's disease. Huffington Post. June 6, 2016. www.huffpost.com/entry/one-of -muhammad-alis-greatest-legacies-is-his-fight-for-people-with-parkinsons -disease_n_5755bfe8e4b0eb20fa0ea112. Accessed October 15, 2018.

2. Geiger LE. Etiologies of parkinsonism. *Bulletin of the Los Angeles Neurological Society.* 1975;40(4):160–194.

3. Bower JH, Maraganore DM, Peterson BJ, McDonnell SK, Ahlskog JE, Rocca WA. Head trauma preceding PD: a case-control study. *Neurology.* 2003;60(10):1610–1615; Siavash J, Mahyar E, Farhad A, Ali S. Head injury and risk of Parkinson disease: a systematic review and meta-analysis. *Movement Disorders.* 2013;28(9):1222–1229; Harris MA, Shen H, Marion SA,

Tsui JK, Teschke K. Head injuries and Parkinson's disease in a case-control study. *Occupational and Environmental Medicine.* 2013;70(12):839–844.

4. Goldman SM, Tanner CM, Oakes D, Bhudhikanok GS, Gupta A, Langston JW. Head injury and Parkinson's disease risk in twins. *Annals of Neurology.* 2006;60(1):65–72.

5. Ibid.

6. Crane PK, Gibbons LE, Dams-O'Connor K, et al. Association of traumatic brain injury with late-life neurodegenerative conditions and neuropathologic findings. *JAMA Neurology.* 2016;73(9):1062–1069; Uryu K, Chen X-H, Martinez D, et al. Multiple proteins implicated in neurodegenerative diseases accumulate in axons after brain trauma in humans. *Experimental Neurology.* 2007;208(2):185–192.

7. Lee P-C, Bordelon Y, Bronstein J, Ritz B. Traumatic brain injury, paraquat exposure, and their relationship to Parkinson disease. *Neurology.* 2012;79(20):2061–2066.

8. Omalu BI, DeKosky ST, Minster RL, Kamboh MI, Hamilton RL, Wecht CH. Chronic traumatic encephalopathy in a National Football League player. *Neurosurgery.* 2005;57(1):128–134; Mez J, Daneshvar DH, Kiernan PT, et al. Clinicopathological evaluation of chronic traumatic encephalopathy in players of American football. *JAMA.* 2017;318(4):360–370; Lehman EJ, Hein MJ, Baron SL, Gersic CM. Neurodegenerative causes of death among retired National Football League players. *Neurology.* 2012;79(19):1970–1974; Kubilus CA, Nowinski CJ, Budson AE, et al. Chronic traumatic encephalopathy in athletes: progressive tauopathy after repetitive head injury. *Journal of Neuropathology & Experimental Neurology.* 2009;68(7):709–735.

9. Lehman EJ, Hein MJ, Baron SL, Gersic CM. Neurodegenerative causes of death among retired National Football League players. *Neurology.* 2012;79(19):1970–1974.

10. Mez J, Daneshvar DH, Kiernan PT, et al. Clinicopathological evaluation of chronic traumatic encephalopathy in players of American football. *JAMA.* 2017;318(4):360–370.

11. Ibid.

12. Borland C. I quit the NFL but have hope for football's future. Huffington Post. February 1, 2018. www.huffingtonpost.com/entry/opinion-borland-nfl-cte_us_5a72024de4b03699143ec7a3. Accessed October 15, 2018.

13. Mez J, Daneshvar DH, Kiernan PT, et al. Clinicopathological evaluation of chronic traumatic encephalopathy in players of American football. *JAMA.* 2017;318(4):360–370.

14. Borland C. I quit the NFL but have hope for football's future. Huffington Post. February 1, 2018. www.huffingtonpost.com/entry/opinion-borland -nfl-cte_us_5a72024de4b03699143ec7a3. Accessed October 15, 2018.

15. Ezell L. Timeline: the NFL's concussion crisis. PBS Frontline. October 8, 2013. www.pbs.org/wgbh/pages/frontline/sports/league-of-denial /timeline-the-nfls-concussion-crisis/#2001. Accessed October 15, 2018.

16. Belson K. Brain trauma to affect one in three players, N.F.L. agrees. New York Times. September 13, 2014:A1.

17. Concussions in American football. Wikipedia. https://en.wikipedia .org/wiki/Concussions_in_American_football#Federal_NFL_concussion _litigation. Accessed October 15, 2018.

18. Hruby P. Startling jump in NFL player claims for Parkinson's and ALS pushes payout projections past 65-year total in 18 months. Los Angeles Times. August 8, 2018.

19. Ibid.

20. Press A. Forrest Gregg fighting Parkinson's. ESPN. November 16, 2011. www.espn.com/nfl/story/_/id/7242215/forrest-gregg-says-fighting -parkinsons. Accessed December 7, 2018.

21. Ibid.

22. ESPN News Services. Hall of Fame lineman Forrest Gregg dies at 85. ESPN. April 12, 2019. www.espn.com/nfl/story/_/id/26505006/hall-fame -lineman-forrest-gregg-dies-85. Accessed May 19, 2019.

23. Kontos AP, Elbin RJ, Fazio-Sumrock VC, et al. Incidence of sports-related concussion among youth football players aged 8–12 years. Journal of Pediatrics. 2013;163(3):717–720.

24. Marar M, McIlvain NM, Fields SK, Comstock RD. Epidemiology of concussions among United States high school athletes in 20 sports. American Journal of Sports Medicine. 2012;40(4):747–755; Gessel LM, Fields SK, Collins CL, Dick RW, Comstock RD. Concussions among United States high school and collegiate athletes. Journal of Athletic Training. 2007;42(4):495–503; Dompier TP, Kerr ZY, Marshall SW, et al. Incidence of concussion during practice and games in youth, high school, and collegiate American football players. JAMA Pediatrics. 2015;169(7):659–665.

25. McCrea M, Hammeke T, Olsen G, Leo P, Guskiewicz K. Unreported concussion in high school football players: implications for prevention. Clinical Journal of Sport Medicine. 2004;14(1):13–17.

26. Kroshus E, Garnett B, Hawrilenko M, Baugh CM, Calzo JP. Concussion under-reporting and pressure from coaches, teammates, fans, and parents. Social Science & Medicine. 2015;134:66–75.

27. Marar M, McIlvain NM, Fields SK, Comstock RD. Epidemiology of concussions among United States high school athletes in 20 sports. *American Journal of Sports Medicine*. 2012;40(1):747–755.

28. Ibid.

29. Ibid.

30. DoD worldwide numbers for TBI. Defense and Veterans Brain Injury Center. https://dvbic.dcoe.mil/dod-worldwide-numbers-tbi. Accessed March 26, 2019.

31. Gardner RC, Byers AL, Barnes DE, Li Y, Boscardin J, Yaffe K. Mild TBI and risk of Parkinson disease: a chronic effects of neurotrauma consortium study. *Neurology*. 2018;90(20):e1771–e1779.

32. DoD worldwide numbers for TBI. Defense and Veterans Brain Injury Center. https://dvbic.dcoe.mil/dod-worldwide-numbers-tbi. Accessed March 26, 2019.

33. DoD worldwide numbers for TBI. Defense and Veterans Brain Injury Center. https://dvbic.dcoe.mil/dod-worldwide-numbers-tbi. Accessed March 26, 2019.

34. Gardner RC, Byers AL, Barnes DE, Li Y, Boscardin J, Yaffe K. Mild TBI and risk of Parkinson disease: a chronic effects of neurotrauma consortium study. *Neurology*. 2018;90(20):e1771–e1779.

35. Ibid.

36. Williams Gowers (neurologist). Wikipedia. https://en.wikipedia.org/wiki/William_Gowers_(neurologist). Accessed March 14, 2019.

37. Ibid.; Gowers W. *Paralysis agitans*. Macmillan; 1899.

38. McCarrick K. Changing course: Jimmy Choi's story. *FoxFeed Blog*. May 29, 2018. www.michaeljfox.org/news/changing-course-jimmy-chois-story. Accessed June 27, 2019.

39. Boiles A. WATCH NOW: athlete with Parkinson's Jimmy Choi becomes an American ninja warrior. *Foxfeed Blog*. July 4, 2017. www.michaeljfox.org/news/watch-now-athlete-parkinsons-jimmy-choi-becomes-american-ninja-warrior. Accessed June 27, 2019.

40. McCarrick K. Changing course: Jimmy Choi's story. *Foxfeed Blog*. May 29, 2018. www.michaeljfox.org/news/changing-course-jimmy-chois-story. Accessed June 27, 2019.

41. Terraso D. Professor pedals across Iowa to test benefits on Parkinson's disease. Georgia Institute of Technology News Center. July 21, 2004. www.news.gatech.edu/2004/07/21/professor-pedals-across-iowa-test-benefits-parkinsons-disease. Accessed March 14, 2019.

42. Bridges A. Pedaling away from Parkinson's. *Santa Barbara Independent*. August 11, 2014.

43. Reynolds G. What Parkinson's teaches us about the brain. Well (blog), New York Times. October 12, 2011. https://well.blogs.nytimes.com/2011/10/12/what-parkinsons-teaches-us-about-the-brain. Accessed June 27, 2019.

44. Pedaling for Parkinson's. www.pedalingforparkinsons.org. Accessed March 14, 2019.

45. Bridges A. Pedaling away from Parkinson's. Santa Barbara Independent. August 11, 2014; Pedaling for Parkinson's. www.pedalingforparkinsons.org. Accessed March 14, 2019.

46. Ridgel AL, Vitek JL, Alberts JL. Forced, not voluntary, exercise improves motor function in Parkinson's disease patients. Neurorehabilitation and Neural Repair. 2009;23(6):600–608.

47. Alberts JL, Linder SM, Penko AL, Lowe MJ, Phillips M. It is not about the bike, it is about the pedaling: forced exercise and Parkinson's disease. Exercise and Sport Sciences Reviews. 2011;39(4):177–186.

48. Ibid.

49. Goodwin VA, Richards SH, Taylor RS, Taylor AH, Campbell JL. The effectiveness of exercise interventions for people with Parkinson's disease: a systematic review and meta-analysis. Movement Disorders. 2008;23(5):631–640.

50. Müller J, Myers J. Association between physical fitness, cardiovascular risk factors, and Parkinson's disease. European Journal of Preventive Cardiology. 2018;25(13):1409–1415.

51. Fang X, Han D, Cheng Q, et al. Association of levels of physical activity with risk of Parkinson disease: a systematic review and meta-analysis. JAMA Network Open. 2018;1(5):e182421; Thacker EL, Chen H, Patel AV, et al. Recreational physical activity and risk of Parkinson's disease. Movement Disorders. 2008;23(1):69–74.

52. Ascherio A, Schwarzschild MA. The epidemiology of Parkinson's disease: risk factors and prevention. Lancet Neurology. 2016;15(12):1257–1272; Sun Q, Townsend MK, Okereke OI, Franco OH, Hu FB, Grodstein F. Physical activity at midlife in relation to successful survival in women at age 70 years or older. Archives of Internal Medicine. 2010;170(2):194–201; Myers J, Prakash M, Froelicher V, Do D, Partington S, Atwood JE. Exercise capacity and mortality among men referred for exercise testing. New England Journal of Medicine. 2002;346(11):793–801; Gebel K, Ding D, Chey T, Stamatakis E, Brown WJ, Bauman AE. Effect of moderate to vigorous physical activity on all-cause mortality in middle-aged and older Australians. JAMA Internal Medicine. 2015;175(6):970–977.

53. Fang X, Han D, Cheng Q, et al. Association of levels of physical activity with risk of Parkinson disease: a systematic review and meta-analysis.

JAMA Network Open. 2018;1(5):e182421; Metabolic equivalent. Wikipedia. https://en.wikipedia.org/wiki/Metabolic_equivalent. Accessed October 15, 2018.

54. Serra-Majem L, Roman B, Estruch R. Scientific evidence of interventions using the Mediterranean diet: a systemic review. *Nutritional Review*. 2006;64:S27–S47; Casini A, Gensini GF, Abbate R, Sofi F. Accruing evidence on benefits of adherence to the Mediterranean diet on health: an updated systematic review and meta-analysis. *American Journal of Clinical Nutrition*. 2010;92(5):1189–1196.

55. Casini A, Gensini GF, Abbate R, Sofi F. Accruing evidence on benefits of adherence to the Mediterranean diet on health: an updated systematic review and meta-analysis. *American Journal of Clinical Nutrition*. 2010;92(5):1189–1196.

56. Alcalay RN, Gu Y, Mejia-Santana H, Cote L, Marder KS, Scarmeas N. The association between Mediterranean diet adherence and Parkinson's disease. *Movement Disorders*. 2012;27(6):771–774.

57. Gao X, Chen H, Fung TT, et al. Prospective study of dietary pattern and risk of Parkinson disease. *American Journal of Clinical Nutrition*. 2007;86(5):1486–1494.

58. Alcalay RN, Gu Y, Mejia-Santana H, Cote L, Marder KS, Scarmeas N. The association between Mediterranean diet adherence and Parkinson's disease. *Movement Disorders*. 2012;27(6):771–774; Sofi F, Cesari F, Abbate R, Gensini GF, Casini A. Adherence to Mediterranean diet and health status: meta-analysis. *BMJ*. 2008;337.

59. Alcalay RN, Gu Y, Mejia-Santana H, Cote L, Marder KS, Scarmeas N. The association between Mediterranean diet adherence and Parkinson's disease. *Movement Disorders*. 2012;27(6):771–774.

60. Gao X, Chen H, Fung TT, et al. Prospective study of dietary pattern and risk of Parkinson disease. *American Journal of Clinical Nutrition*. 2007;86(5):1486–1494.

61. Jin H, Kanthasamy A, Ghosh A, Anantharam V, Kalyanaraman B, Kanthasamy AG. Mitochondria-targeted antioxidants for treatment of Parkinson's disease: preclinical and clinical outcomes. *Biochimica et Biophysica Acta*. 2014;1842(8):1282–1294.

62. Ross G, Abbott RD, Petrovitch H, et al. Association of coffee and caffeine intake with the risk of Parkinson disease. *JAMA*. 2000;283(20):2674–2679; Paganini-Hill A. Risk factors for Parkinson's disease: the leisure world cohort study. *Neuroepidemiology*. 2001;20(2):118–124; Ascherio A, Zhang SM, Hernan MA, et al. Prospective study of caffeine consumption and risk of

Parkinson's disease in men and women. *Annals of Neurology.* 2001;50(1):56–63; Hernan MA, Takkouche B, Caamano-Isorna F, Gestal-Otero JJ. A meta-analysis of coffee drinking, cigarette smoking, and the risk of Parkinson's disease. *Annals of Neurology.* 2002;52(3):276–284; Tan LC, Koh WP, Yuan JM, et al. Differential effects of black versus green tea on risk of Parkinson's disease in the Singapore Chinese Health Study. *American Journal of Epidemiology.* 2008;167(5):553–560; Liu R, Guo X, Park Y, et al. Caffeine intake, smoking, and risk of Parkinson disease in men and women. *American Journal of Epidemiology.* 2012;175(11):1200–1207.

63. Ross G, Abbott RD, Petrovitch H, et al. Association of coffee and caffeine intake with the risk of Parkinson disease. *JAMA.* 2000;283(20):2674–2679.

64. Ibid.; Munoz DG, Fujioka S. Caffeine and Parkinson disease: a possible diagnostic and pathogenic breakthrough. *Neurology.* 2018;90(5):205–206.

65. Ascherio A, Zhang SM, Hernan MA, et al. Prospective study of caffeine consumption and risk of Parkinson's disease in men and women. *Annals of Neurology.* 2001;50(1):56–63.

66. Munoz DG, Fujioka S. Caffeine and Parkinson disease: A possible diagnostic and pathogenic breakthrough. *Neurology.* 2018;90(5):205–206; Schwarzschild MA, Chen JF, Ascherio A. Caffeinated clues and the promise of adenosine A(2A) antagonists in PD. *Neurology.* 2002;58(8):1154–1160.

67. Postuma RB, Anang J, Pelletier A, et al. Caffeine as symptomatic treatment for Parkinson disease (Café-PD): a randomized trial. *Neurology.* 2017;89(17):1795–1803.

68. Palacios N, Gao X, McCullough ML, et al. Caffeine and risk of Parkinson's disease in a large cohort of men and women. *Movement Disorders.* 2012;27(10):1276–1282.

69. Wikoff D, Welsh BT, Henderson R, et al. Systematic review of the potential adverse effects of caffeine consumption in healthy adults, pregnant women, adolescents, and children. *Food and Chemical Toxicology.* 2017;109:585–648.

CHAPTER 7: TAKING CARE

1. Boston JG. JCC Greater Boston—building a new beginning [video file]. YouTube. March 23, 2018. www.youtube.com/watch?v=unw3Ue2 DWME.

2. Lagro-Janssen ALM, van den Heuvel EAJ, Bor HH, et al. Prodromal symptoms and early detection of Parkinson's disease in general practice: a nested case-control study. *Family Practice.* 2014;31(4):373–378; Gonera EG, Hof MVt, Berger HJC, van Weel C, Horstink MWIM. Symptoms and

duration of the prodromal phase in Parkinson's disease. *Movement Disorders.* 1997;12(6):871–876.

3. Osler W, Sir. *The Principles and Practice of Medicine.* D. Appleton and Company; 1892.

4. Taylor KSM, Counsell CE, Harris CE, Gordon JC. Screening for undiagnosed parkinsonism in people aged 65 years and over in the community. *Parkinsonism & Related Disorders.* 2006;12(2):79–85.

5. Nicoletti A, Sofia V, Bartoloni A, et al. Prevalence of Parkinson's disease: a door-to-door survey in rural Bolivia. *Parkinsonism & Related Disorders.* 2003;10(1):19–21.

6. Kis B, Schrag A, Ben-Shlomo Y, et al. Novel three-stage ascertainment method: prevalence of PD and parkinsonism in South Tyrol, Italy. *Neurology.* 2002;58(12):1820–1825.

7. Muangpaisan W, Hori H, Brayne C. Systematic review of the prevalence and incidence of Parkinson's disease in Asia. *Journal of Epidemiology.* 2009;19(6):281–293.

8. Schoenberg BS, Anderson DW, Haerer AF. Prevalence of Parkinson's disease in the biracial population of Copiah County, Mississippi. *Neurology.* 1985;35(6):841–845.

9. Safarpour D, Thibault DP, DeSanto CL, et al. Nursing home and end-of-life care in Parkinson disease. *Neurology.* 2015;85(5):413–419.

10. Buchanan RJ, Wang S, Huang C, Simpson P, Manyam BV. Analyses of nursing home residents with Parkinson's disease using the minimum data set. *Parkinsonism & Related Disorders.* 2002;8(5):369–380.

11. Weerkamp NJ, Zuidema SU, Tissingh G, et al. Motor profile and drug treatment of nursing home residents with Parkinson's disease. *Journal of the American Geriatrics Society.* 2012;60(12):2277–2282.

12. Safarpour D, Thibault DP, DeSanto CL, et al. Nursing home and end-of-life care in Parkinson disease. *Neurology.* 2015;85(5):413–419.

13. Weerkamp NJ, Tissingh G, Poels PJE, et al. Diagnostic accuracy of Parkinson's disease and atypical parkinsonism in nursing homes. *Parkinsonism & Related Disorders.* 2014;20(11):1157–1160.

14. Willis AW, Schootman M, Evanoff BA, Perlmutter JS, Racette BA. Neurologist care in Parkinson disease: a utilization, outcomes, and survival study. *Neurology.* 2011;77(9):851–857.

15. Ibid.

16. Ibid.

17. Papanicolas I, Woskie LR, Jha AK. Health care spending in the United States and other high-income countries. *JAMA.* 2018;319(10):1024–1039;

Dahodwala N, Xie M, Noll E, Siderowf A, Mandell DS. Treatment dispari-ties in Parkinson's disease. *Annals of Neurology.* 2009;66(2):142–145.

18. Willis AW, Schootman M, Evanoff BA, Perlmutter JS, Racette BA. Neurologist care in Parkinson disease: a utilization, outcomes, and survival study. *Neurology.* 2011;77(9):851–857.

19. Willis AW, Schootman M, Kung N, Wang X-Y, Perlmutter JS, Racette BA. Disparities in deep brain stimulation surgery among insured elders with Parkinson disease. *Neurology.* 2014;82(2):163–171.

20. Willis AW, Schootman M, Tran R, et al. Neurologist-associated reduc-tion in PD-related hospitalizations and health care expenditures. *Neurology.* 2012;79(17):1774–1780.

21. Gage H, Hendricks A, Zhang S, Kazis L. The relative health related quality of life of veterans with Parkinson's disease. *Journal of Neurology, Neu-rosurgery, and Psychiatry.* 2003;74(2):163–169.

22. Seidman C. Caring for wife with dementia leads husband to contem-plate suicide. *Sarasota Herald-Tribune.* July 12, 2018; Opinion.

23. Caregiving in the U.S. 2015—executive summary. National Alliance for Caregiving. June 2015. www.caregiving.org/wp-content/uploads/2015/05/2015_CaregivingintheUS_Executive-Summary-June-4_WEB.pdf. Accessed June 27, 2019.

24. Schrag A, Hovris A, Morley D, Quinn N, Jahanshahi M. Caregiver-burden in Parkinson's disease is closely associated with psychiatric symptoms, falls, and disability. *Parkinsonism & Related Disorders.* 2006;12(1):35–41.

25. Schulz R, Beach SR. Caregiving as a risk factor for mortality: the care-giver health effects study. *JAMA.* 1999;282(23):2215–2219.

26. Ahlskog JE. Beating a dead horse: dopamine and Parkinson disease. *Neurology.* 2007;69(17):1701–1711.

27. Nijkrake MJ, Keus SHJ, Overeem S, et al. The ParkinsonNet con-cept: development, implementation and initial experience. *Movement Dis-orders.* 2010;25(7):823–829; Bloem BR, Rompen L, de Vries NM, Klink A, Munneke M, Jeurissen P. ParkinsonNet: A low-cost health care in-novation with a systems approach from the Netherlands. *Health Affairs.* 2017;36:1987–1996; Munneke M, Nijkrake MJ, Keus SHJ, et al. Efficacy of community-based physiotherapy networks for patients with Parkinson's disease: a cluster-randomised trial. *Lancet Neurology.* 2010;9(1):46–54; Wensing M, van der Eijk M, Koetsenruijter J, Bloem BR, Munneke M, Faber M. Connectedness of healthcare professionals involved in the treat-ment of patients with Parkinson's disease: a social networks study. *Imple-mentation Science.* 2011;6(1):67; Keus SHJ, Oude Nijhuis LB, Nijkrake

MJ, Bloem BR, Munneke M. Improving community healthcare for patients with Parkinson's disease: the Dutch model. *Journal of Parkinson's Disease.* 2012;2012:543426; van der Ejik M, Faber MJ, Aarts JW, Kremer JA, Munneke M, Bloem BR. Using online health communities to deliver patient-centered care to people with chronic conditions. *Journal of Medical Internet Research.* 2013;15(6):e115; Sturkenboom IHWM, Graff MJL, Hendriks JCM, et al. Efficacy of occupational therapy for patients with Parkinson's disease: a randomised controlled trial. *Lancet Neurology.* 2014;13(6):557–566; Bloem BR, Munneke M. Revolutionising management of chronic disease: the ParkinsonNet approach. *BMJ.* 2014;348:g1838; Ypinga JHL, de Vries NM, Boonen LHHM, et al. Effectiveness and costs of specialised physiotherapy given via ParkinsonNet: a retrospective analysis of medical claims data. *Lancet Neurology.* 2018;17(2):153–161.

28. Bloem BR, Rompen L, de Vries NM, Klink A, Munneke M, Jeurissen P. ParkinsonNet: a low-cost health care innovation with a systems approach from the Netherlands. *Health Affairs.* 2017;36:1987–1996.

29. Ibid.

30. Dorsey ER, Glidden AM, Holloway MR, Birbeck GL, Schwamm LH. Teleneurology and mobile technologies: the future of neurological care. *Nature Reviews Neurology.* 2018;14:285–297.

31. Arora S, Thornton K, Murata G, et al. Outcomes of Treatment for Hepatitis C Virus Infection by Primary Care Providers. *New England Journal of Medicine.* 2011;364(23):2199–2207.

32. Arora S. Project ECHO: the evidence is catching up with the enthusiasm. *Health Affairs Blog.* January 13, 2017. www.healthaffairs.org/do/10.1377/hblog20170113.058331/full. Accessed June 27, 2019; Catic AG, Mattison MLP, Bakaev I, Morgan M, Monti SM, Lipsitz L. ECHO-AGE: an innovative model of geriatric care for long-term care residents with dementia and behavioral issues. *Journal of the American Medical Directors Association.* 2014;15(12):938–942; Project ECHO: a revolution in medical education and care delivery. University of New Mexico. https://echo.unm.edu. Accessed November 11, 2018.

33. Dorsey ER, Glidden AM, Holloway MR, Birbeck GL, Schwamm LH. Teleneurology and mobile technologies: the future of neurological care. *Nature Reviews Neurology.* 2018;14:285–297.

34. Meyer GS, Gibbons RV. House calls to the elderly—a vanishing practice among physicians. *New England Journal of Medicine.* 1997;337(25):1815–1820.

35. Ibid.

36. Sairenji T, Jetty A, Peterson LE. Shifting patterns of physician home visits. *Journal of Primary Care & Community Health*. 2015;7(2):71–75.

37. Fleisher J, Barbosa W, Sweeney MM, et al. Interdisciplinary home visits for individuals with advanced Parkinson's disease and related disorders. *Journal of the American Geriatrics Society*. 2018;66(6):1226–1232.

38. Ibid.

39. Hack N, Akbar U, Monari, EH, et al. Person-centered care in the home setting for Parkinson's disease: operation house call quality of care pilot study. *Journal of Parkinson's Disease*. 2015;2015;639494. doi:10.1155/2015/639494. www.ncbi.nlm.nih.gov/pmc/articles/PMC4452493.

40. Beck CA, Beran DB, Biglan KM, et al. National randomized controlled trial of virtual house calls for Parkinson disease. *Neurology*. 2017;89(11):1152–1161; Shah SP, Glenn GL, Hummel EM, et al. Caregiver tele-support group for Parkinson's disease: a pilot study. *Geriatric Nursing*. 2015;36(3):207–211; Korn RE, Shukla AW, Katz M, et al. Virtual visits for Parkinson disease. *Neurology: Clinical Practice*. 2017;7(4):283–295; Venkataraman V, Donohue SJ, Biglan KM, Wicks P, Dorsey ER. Virtual visits for Parkinson disease. *Neurology: Clinical Practice*. 2014;4(2):146–152.

41. Beck CA, Beran DB, Biglan KM, et al. National randomized controlled trial of virtual house calls for Parkinson disease. *Neurology*. 2017;89(11):1152–1161; Dorsey ER, Venkataraman V, Grana MJ, et al. Randomized controlled clinical trial of "virtual house calls" for Parkinson disease. *JAMA Neurology*. 2013;70(5):565–570; Heikkilä VM, Turkka J, Korpelainen J, Kallanranta T, Summala H. Decreased driving ability in people with Parkinson's disease. *Journal of Neurology, Neurosurgery & Psychiatry*. 1998;64(3):325–330.

42. Beck CA, Beran DB, Biglan KM, et al. National randomized controlled trial of virtual house calls for Parkinson disease. *Neurology*. 2017;89(11):1152–1161; Dorsey ER, Venkataraman V, Grana MJ, et al. Randomized controlled clinical trial of "virtual house calls" for Parkinson disease. *JAMA Neurology*. 2013;70(5):565–570; Heikkilä VM, Turkka J, Korpelainen J, Kallanranta T, Summala H. Decreased driving ability in people with Parkinson's disease. *Journal of Neurology, Neurosurgery & Psychiatry*. 1998;64(3):325–330.

43. Beck CA, Beran DB, Biglan KM, et al. National randomized controlled trial of virtual house calls for Parkinson disease. *Neurology*. 2017;89(11):1152–1161.

44. Hubble JP. Interactive video conferencing and Parkinson's disease. *Kansas Medicine*. 1992;93(12):351–352.

45. Ibid.

46. Achey M, Aldred JL, Aljehani N, et al. The past, present, and future of telemedicine for Parkinson's disease. Movement Disorders 2014;29(7):871–883. https://onlinelibrary.wiley.com/doi/full/10.1002/mds.25903.

47. Silva de Lima AL, Hahn T, Evers LJW, et al. Feasibility of large-scale deployment of multiple wearable sensors in Parkinson's disease. PLOS ONE. 2017;12(12):e0189161.

48. Lipsmeier F, Taylor KI, Kilchenmann T, et al. Evaluation of smartphone-based testing to generate exploratory outcome measures in a phase 1 Parkinson's disease clinical trial. Movement Disorders. 2018;33(8):1287–1297.

49. Dorsey ER, Glidden AM, Holloway MR, et al. Teleneurology and mobile technologies: the future of neurological care. Nat Rev Neurol 2018;14(5):285–297. www.ncbi.nlm.nih.gov/pubmed/29623949.

50. Sharma S, Padma MV, Bhardwaj A, Sharma A, Sawal N, Thakur S. Telestroke in resource-poor developing country model. Neurology India. 2016;64(5):934–940; Corley J. Telestroke: India's solution to a public health-care crisis. Lancet Neurology. 2018;17(2):115–116.

51. Topol E. The Patient Will See You Now: The Future of Medicine Is in Your Hands. Rpt. ed. Basic Books; 2016.

52. Daschle T, Dorsey ER. The return of the house call. Annals of Internal Medicine. 2015;162(8):587–588.

53. Gornick ME, Warren JL, Eggers PW, et al. Thirty years of Medicare: impact on the covered population. Health Care Financing Review. 1996;18(2):179–237; Life expectancy for Social Security. Social Security Administration. www.ssa.gov/history/lifeexpect.html. Accessed November 10, 2018.

54. Thorpe KE, Ogden LL, Galactionova K. Chronic conditions account for rise in Medicare spending from 1987 to 2006. Health Affairs. 2010;29(4):718–724.

55. AARP. Baby boomer facts and figures. AARP Livable Communities. www.aarp.org/livable-communities/info-2014/livable-communities-facts-and-figures.html. Accessed January 5, 2019.

56. Dieleman JL, Baral R, Birger M, et al. US spending on personal health care and public health, 1996–2013. JAMA. 2016;316(24):2627–2646.

57. Ibid.

58. Dorsey ER, George BP, Leff B, Willis AW. The coming crisis: obtaining care for the growing burden of neurodegenerative conditions. Neurology. 2013;80(21):1989–1996.

59. Ibid.; Barnett ML, Wilcock A, McWilliams JM, et al. Two-year evaluation of mandatory bundled payments for joint replacement. New En-

gland Journal of Medicine. 2019;380(3):252–262; Silverman L, Walters D, Aasen E, Tainter R, Whittington J, Holter R. Chapter 7: the cost of a broken hip. KERA. 2017. http://stories.kera.org/the-broken-hip/the-cost -of-a-broken-hip. Accessed February 11, 2019.

60. Dorsey ER, Vlaanderen FP, Engelen LJ, et al. Moving Parkinson care to the home. *Movement Disorders.* 2016;31(9):1258–1262.

61. Ibid.; Gerlach OHH, Winogrodzka A, Weber WEJ. Clinical problems in the hospitalized Parkinson's disease patient: systematic review *Movement Disorders.* 2011,26(2):197–208; Aminoff MJ, Christine CW, Friedman JH, et al. Management of the hospitalized patient with Parkinson's disease: current state of the field and need for guidelines. *Parkinsonism & Related Disorders.* 2011;17(3):139–145.

62. Ypinga JHL, de Vries NM, Boonen LHHM, et al. Effectiveness and costs of specialised physiotherapy given via ParkinsonNet: a retrospective analysis of medical claims data. *Lancet Neurology.* 2018;17(2):153–161.

62. Parkinson's disease insight report. Wilmington Healthcare. https:// wilmingtonhealthcare.com/what-we-do/nhs-service-improvement /parkinsons-disease-insight-report. Accessed May 19, 2019.

64. Allen NE, Schwarzel AK, Canning CG. Recurrent falls in Parkinson's disease: a systematic review. *Journal of Parkinson's Disease.* 2013;2013:906274.

65. Walker RW, Chaplin A, Hancock RL, Rutherford R, Gray WK. Hip fractures in people with idiopathic Parkinson's disease: incidence and outcomes. *Movement Disorders.* 2013;28(3):334–340; Kalilani L, Asgharnejad M, Palokangas T, Durgin T. Comparing the incidence of falls/ fractures in Parkinson's disease patients in the US population. *PLOS ONE.* 2016;11(9):e0161689–e0161689.

66. Li F, Harmer P, Fitzgerald K, et al. Tai chi and postural stability in patients with Parkinson's disease. *New England Journal of Medicine.* 2012;366(6):511–519; Ypinga JHL, de Vries NM, Boonen LHHM, et al. Effectiveness and costs of specialised physiotherapy given via ParkinsonNet: a retrospective analysis of medical claims data. *Lancet Neurology.* 2018;17(2):153–161; van der Marck MA, Klok MPC, Okun MS, Giladi N, Munneke M, Bloem BR. Consensus-based clinical practice recommendations for the examination and management of falls in patients with Parkinson's disease. *Parkinsonism & Related Disorders.* 2014;20(4):360–369.

67. Florence CS, Bergen G, Atherly A, Burns E, Stevens J, Drake C. Medical costs of fatal and nonfatal falls in older adults. *Journal of the American Geriatrics Society.* 2018;66(4):693–698.

68. Ibid.; Kaiser J. Final 2018 budget bill eases biomedical researchers' policy worries. *Science*. March 22, 2018. American Association for the Advancement of Science. www.sciencemag.org/news/2018/03/final-2018-budget-bill-eases-biomedical-researchers-policy-worries. Accessed June 27, 2019.

69. Safarpour D, Thibault DP, DeSanto CL, et al. Nursing home and end-of-life care in Parkinson disease. *Neurology*. 2015;85(5):413–419.

70. Dorsey ER, George BP, Leff B, Willis AW. The coming crisis: obtaining care for the growing burden of neurodegenerative conditions. *Neurology*. 2013;80(21):1989–1996; Powell T. New hopes for dementia care. *Wall Street Journal*. April 11, 2019.

71. Dorsey ER, Topol EJ. State of telehealth. *NEJM*. 2016;375(2):154–161.

72. Wilkinson JR, Spindler M, Wood SM, et al. High patient satisfaction with telehealth in Parkinson disease: a randomized controlled study. *Neurology Clinical Practice*. 2016;6(3):241–251.

73. Telehealth services. In: *Medicare Learning Network Booklet*. Centers for Medicare & Medicaid Services; 2018.

74. Lacktman NM. Medicare payments for telehealth increased 28% in 2016: what you should know. Health Care Law Today. August 28, 2017. www.healthcarelawtoday.com/2017/08/28/medicare-payments-for-telehealth-increased-28-in-2016-what-you-should-know. Accessed November 2018.

75. Dorsey ER, George BP, Leff B, Willis AW. The coming crisis: obtaining care for the growing burden of neurodegenerative conditions. *Neurology*. 2013;80(21):1989–1996.

76. Rosenthal E. *An American Sickness: How Healthcare Became Big Business and How You Can Take It Back*. 1st ed. Penguin Press; 2017; Goodson JD. Unintended consequences of resource-based relative value scale reimbursement. *JAMA*. 2007;298(19):2308–2310.

77. Rosenthal E. *An American Sickness: How Healthcare Became Big Business and How You Can Take It Back*. 1st ed. Penguin Press; 2017.

78. Composition of the RVS Update Committee (RUC). American Medical Association. www.ama-assn.org/about/rvs-update-committee-ruc/composition-rvs-update-committee-ruc. Accessed March 26, 2019.

79. Ibid.

80. Ibid.; Struhal W, Sellner J, Lisnic V, Vécsei L, Müller E, Grisold W. Neurology residency training in Europe—the current situation. *European Journal of Neurology*. 2011;18(4):e36–e40.

81. Rosenthal E. *An American Sickness: How Healthcare Became Big Business and How You Can Take It Back*. 1st ed. Penguin Press; 2017.

82. Dorsey ER, George BP, Leff B, Willis AW. The coming crisis: obtaining care for the growing burden of neurodegenerative conditions. *Neurology*. 2013;80(21):1989–1996.

83. FAQs about the FDA Patient Representative Program. US Food and Drug Administration. www.fda.gov/ForPatients/ucm412529.htm. Updated May 12, 2018. Accessed March 19, 2019.

84. Dorsey ER, George BP, Leff B, Willis AW. The coming crisis: obtaining care for the growing burden of neurodegenerative conditions. *Neurology*, 2013;80(21):1989–1996; Rosenthal E. *An American Sickness: How Healthcare Became Big Business and How You Can Take It Back*. 1st ed. Penguin Press; 2017.

85. Historical evolution of FDA advisory committees. In: Rettig RA, Earley LE, Merill RA, eds. *Institute of Medicine (US) Committee to Study the Use of Advisory Committees*. National Academies Press; 1992.

CHAPTER 8: HOPE ON THE HORIZON

1. FDA approved drugs for neurology. CenterWatch. www.centerwatch .com/drug-information/fda-approved-drugs/therapeutic-area/10/neurology. Accessed November 29, 2018.

2. Ibid.

3. Prescribing information: INBRIJA. US Food and Drug Administration. December 2018. www.accessdata.fda.gov/drugsatfda_docs/label/2018 /209184s000lbl.pdf.

4. Prescribing information: GOCOVRI. US Food and Drug Administration. August 2017. www.accessdata.fda.gov/drugsatfda_docs/label/2017 /208944lbl.pdf.

5. Prescribing information: XADAGO. US Food and Drug Administration. March 2017. www.accessdata.fda.gov/drugsatfda_docs/label/2017 /207145lbl.pdf.

6. Prescribing information: NUPLAZID. US Food and Drug Administration. April 2016. www.accessdata.fda.gov/drugsatfda_docs/label/2016 /207318lbl.pdf.

7. Prescribing information: DUOPA. US Food and Drug Administration. September 2016. www.rxabbvie.com/pdf/duopa_pi.pdf.

8. Prescribing information: RYTARY. US Food and Drug Administration. January 2015. www.accessdata.fda.gov/drugsatfda_docs/label/2015 /203312s000lbl.pdf.

9. Prescribing information: NORTHERA. US Food and Drug Administration. February 2014. www.accessdata.fda.gov/drugsatfda_docs/label/2014/203202lbl.pdf.

10. Novartis. Prescribing information: EXELON. US Food and Drug Administration. December 2018. www.pharma.us.novartis.com/sites/www.pharma.us.novartis.com/files/exelon.pdf.

11. Prescribing information: NEUPRO. US Food and Drug Administration. April 2012. www.accessdata.fda.gov/drugsatfda_docs/label/2012/021829s001lbl.pdf.

12. Prescribing information: APOKYN. US Food and Drug Administration. July 2014. www.accessdata.fda.gov/drugsatfda_docs/label/2014/021264s010lbl.pdf.

13. Novartis. Prescribing information: COMTAN. US Food and Drug Administration. September 28, 2010. www.accessdata.fda.gov/drugsatfda_docs/label/2010/020796s15lbl.pdf.

14. FDA approved drugs for neurology. CenterWatch. www.centerwatch.com/drug-information/fda-approved-drugs/therapeutic-area/10/neurology. Accessed November 29, 2018.

15. Deuschl G, Schade-Brittinger C, Krack P, et al. A randomized trial of deep-brain stimulation for Parkinson's disease. *New England Journal of Medicine.* 2006;355(9):896–908.

16. Goetz CG. The history of Parkinson's disease: early clinical descriptions and neurological therapies. *Cold Spring Harbor Perspectives in Medicine.* 2011;1(1):a008862; Wagle Shukla A, Okun MS. Surgical treatment of Parkinson's disease: patients, targets, devices, and approaches. *Neurotherapeutics.* 2014;11(1):47–59.

17. Wagle Shukla A, Okun MS. Surgical treatment of Parkinson's disease: patients, targets, devices, and approaches. *Neurotherapeutics.* 2014;11(1):47–59.

18. Alexander GE, Crutcher MD, DeLong MR. Basal ganglia-thalamocortical circuits: parallel substrates for motor, oculomotor, "prefrontal" and "limbic" functions. In: Uylings HBM, Van Eden CG, De Bruin JPC, Corner MA, Feenstra MGP, eds. *Progress in Brain Research.* Vol 85. Elsevier; 1991:119–146; Alexander GE, DeLong MR, Strick PL. Parallel organization of functionally segregated circuits linking basal ganglia and cortex. *Annual Review of Neuroscience.* 1986;9(1):357–381.

19. Benabid AL, Pollak P, Louveau A, Henry S, de Rougemont J. Combined (thalamotomy and stimulation) stereotactic surgery of the VIM thalamic nucleus for bilateral Parkinson disease. *Stereotactic and Functional Neurosurgery.* 1987;50(1–6):344–346.

20. Lozano AM, Eltahawy H. How does DBS work? *Supplement to Clinical Neurophysiology.* 2004;57:733–736.

21. Deuschl G, Schade-Brittinger C, Krack P, et al. A randomized trial of deep-brain stimulation for Parkinson's disease. *New England Journal of Medicine.* 2006;355(9):896–908; Follett KA, Weaver FM, Stern M, et al. Pallidal versus subthalamic deep-brain stimulation for Parkinson's disease. *New England Journal of Medicine.* 2010;362(22):2077–2091; Benabid AL, Chabardes S, Mitrofanis J, Pollak P. Deep brain stimulation of the subthalamic nucleus for the treatment of Parkinson's disease. *Lancet Neurology.* 2009;8(1):67–81; Deep-Brain Stimulation for Parkinson's Disease Study Group. Deep-brain stimulation of the subthalamic nucleus or the pars interna of the globus pallidus in Parkinson's disease. *New England Journal of Medicine.* 2001;345(13):956–963; Weaver FM, Follett K, Stern M, et al. Bilateral deep brain stimulation vs best medical therapy for patients with advanced Parkinson disease: a randomized controlled trial. *JAMA.* 2009;301(1):63–73; Bronstein JM, Tagliati M, Alterman RL, et al. Deep brain stimulation for Parkinson disease: an expert consensus and review of key issues. *Archives of Neurology.* 2011;68(2):165.

22. Limousin P, Foltynie T. Long-term outcomes of deep brain stimulation in Parkinson disease. *Nature Reviews Neurology.* 2019;15(4):234–242.

23. Medtronic Activa Tremor Control System. US Food and Drug Administration. Premarket Approval (PMA). www.accessdata.fda.gov/scripts/cdrh/cfdocs/cfPMA/pma.cfm?id=P960009. Updated March 25, 2019. Accessed March 29, 2019.

24. Rossi PJ, Giordano J, Okun MS. The problem of funding off-label deep brain stimulation: bait-and-switch tactics and the need for policy reform. *JAMA Neurology.* 2017;74(1):9–10; Abosch A, Timmermann L, Bartley S, et al. An international survey of deep brain stimulation procedural steps. *Stereotactic and Functional Neurosurgery.* 2013;91(1):1–11.

25. Eberle W, Penders J, Yazicioglu RF. Closing the loop for deep brain stimulation implants enables personalized healthcare for Parkinson's disease patients. Paper presented at the 2011 Annual International Conference of the IEEE Engineering in Medicine and Biology Society, August 30–September 3, 2011; Patel S, Mancinelli C, Hughes R, Dalton A, Shih L, Bonato P. Optimizing deep brain stimulation settings using wearable sensing technology. Paper presented at the 2009 4th International IEEE/EMBS Conference on Neural Engineering, April 29–May 2, 2009.

26. Anidi C, O'Day JJ, Anderson RW, et al. Neuromodulation targets pathological not physiological beta bursts during gait in Parkinson's disease. *Neurobiology of Disease.* 2018;120:107–117; Malekmohammadi M, Herron J,

Velisar A, et al. Kinematic adaptive deep brain stimulation for resting tremor in Parkinson's disease. *Movement Disorders.* 2016;31(3):426–428.

27. Parker KL, Kim Y, Alberico SL, Emmons EB, Narayanan NS. Optogenetic approaches to evaluate striatal function in animal models of Parkinson disease. *Dialogues in Clinical Neuroscience.* 2016;18(1):99–107.

28. Henderson R. The purple membrane from *Halobacterium halobium. Annual Review of Biophysics and Bioengineering.* 1977;6:87–109.

29. Ibid.

30. Parker KL, Kim Y, Alberico SL, Emmons EB, Narayanan NS. Optogenetic approaches to evaluate striatal function in animal models of Parkinson disease. *Dialogues in Clinical Neuroscience.* 2016;18(1):99–107.

31. Ibid.; Fenno L, Yizhar O, Deisseroth K. The development and application of optogenetics. *Annual Review of Neuroscience.* 2011;34(1):389–412.

32. Kravitz AV, Freeze BS, Parker PRL, et al. Regulation of parkinsonian motor behaviours by optogenetic control of basal ganglia circuitry. *Nature.* 2010;466(7306):622–626.

33. Ibid.

34. Ibid.

35. Alessi DR, Sammler E. LRRK2 kinase in Parkinson's disease. *Science.* 2018;360(6384):36–37.

36. Ibid.

37. Denali Therapeutics announces positive clinical results from LRRK2 inhibitor program for Parkinson's disease. Denali Therapeutics. August 1, 2018. http://investors.denalitherapeutics.com/news-releases/news-release-details/denali-therapeutics-announces-positive-clinical-results-lrrk2#ir-pages. Accessed November 29, 2018.

38. Denali Therapeutics announces first patient dosed in phase 1B study of DNL201 for Parkinson's disease [press release]. Denali Therapeutics, December 10, 2018; Study to evaluate DNL201 in subjects with Parkinson's disease. US National Library of Medicine: ClinicalTrials.gov. https://clinicaltrials.gov/ct2/show/NCT03710707. Updated March 28, 2019. Accessed March 29, 2019.

39. Tracking LRRK2-Parkinson's progression for better trial design. Alzforum. March 23, 2018. www.alzforum.org/news/research-news/tracking-lrrk2-parkinsons-progression-better-trial-design. Accessed November 30, 2018.

40. Alessi DR, Sammler E. LRRK2 kinase in Parkinson's disease. *Science.* 2018;360(6384):36–37.

41. Ibid.; Stott S. New LRRK2 results: game changer? Science of Parkinson's. July 25, 2018. https://scienceofparkinsons.com/2018/07/25/lrrk-2.

Accessed June 27, 2019; Di Maio R, Hoffman EK, Rocha EM, et al. LRRK2 activation in idiopathic Parkinson's disease. *Science Translational Medicine*. 2018;10(451).

42. Di Maio R, Hoffman EK, Rocha EM, et al. LRRK2 activation in idiopathic Parkinson's disease. *Science Translational Medicine*. 2018;10(451).

43. Goker-Alpan O, Schiffmann R, LaMarca ME, Nussbaum RL, McInerney-Leo A, Sidransky E. Parkinsonism among Gaucher disease carriers. *Journal of Medical Genetics*. 2004;41(12):937–940.

44. Klein C, Westenberger A. Genetics of Parkinson's disease. *Cold Spring Harbor Perspectives in Medicine*. 2012;2(1):a008888; Alcalay RN, Caccappolo E, Mejia-Santana H, et al. Frequency of known mutations in early-onset Parkinson disease: implication for genetic counseling: the Consortium on Risk for Early-Onset Parkinson Disease study. *Archives of Neurology*. 2010;67(9):1116–1122.

45. Sidransky E, Lopez G. The link between the GBA gene and parkinsonism. *Lancet Neurology*. 2012;11(11):986–998.

46. Velayati A, Yu WH, Sidransky E. The role of glucocerebrosidase mutations in Parkinson disease and Lewy body disorders. *Current Neurology and Neuroscience Reports*. 2010;10(3):190–198.

47. Konrad A. Meet the top VC in a race to find his own Parkinson's cure. *Forbes*. April 18, 2017. www.forbes.com/sites/alexkonrad/2017/04/18/jonathan-silverstein-venturing-for-a-cure/#4d17b6ba50a3. Accessed June 27, 2019.

48. The Silverstein Foundation in the News. Silverstein Foundation for Parkinson's with GBA. www.silversteinfoundation.org/in-the-news. Accessed March 29, 2019.

49. On the money. CNBC. August 19, 2017. https://archive.org/details/CNBC_20170819_093000_On_the_Money.

50. Silverstein Foundation for Parkinson's with GBA. 2018. www.silversteinfoundation.org. Accessed November 30, 2018.

51. Konrad A. Meet the top VC in a race to find his own Parkinson's cure. *Forbes*. April 18, 2017. www.forbes.com/sites/alexkonrad/2017/04/18/jonathan-silverstein-venturing-for-a-cure/#4d17b6ba50a3. Accessed June 27, 2019.

52. About Prevail. Prevail Therapeutics. www.prevailtherapeutics.com/about-prevail. Accessed November 30, 2018.

53. Silveira CRA, MacKinley J, Coleman K, et al. Ambroxol as a novel disease-modifying treatment for Parkinson's disease dementia: protocol for a single-centre, randomized, double-blind, placebo-controlled trial. *BMC*

Neurology. 2019;19(1):20–20; McNeill A, Magalhaes J, Shen C, et al. Ambroxol improves lysosomal biochemistry in glucocerebrosidase mutation-linked Parkinson disease cells. *Brain.* 2014;137(5):1481–1495; Sardi SP, Cedarbaum JM, Brundin P. Targeted therapies for Parkinson's disease: from genetics to the clinic. *Movement Disorders.* 2018;33(5):684–696.

54. Kuhl MM. First trial begins testing drug in people with GBA mutation. *FoxFeed Blog.* February 14, 2017. www.michaeljfox.org/news/first-trial -begins-testing-drug-people-gba-mutation. Accessed June 27, 2019.

55. Ibid.; Sanofi initiates phase 2 clinical trial to evaluate therapy for genetic form of Parkinson's disease [press release]. Business Wire. February 14, 2017. www.businesswire.com/news/home/20170214005771/en/Sanofi -Initiates-Phase-2-Clinical-Trial-Evaluate. Accessed June 27, 2019; A global study to assess the drug dynamics, efficacy, and safety of GZ/SAR402671 in Parkinson's disease patients carrying a glucocerebrosidase (GBA) gene mutation (MOVES-PD). US National Library of Medicine: ClinicalTrials .gov. https://clinicaltrials.gov/ct2/show/NCT02906020. Updated March 14, 2019. Accessed March 29, 2019; Sardi SP, Viel C, Clarke J, et al. Glucosylceramide synthase inhibition alleviates aberrations in synucleinopathy models. *Proceedings of the National Academy of Sciences of the United States of America.* 2017;114(10):2699–2704.

56. A global study to assess the drug dynamics, efficacy, and safety of GZ/ SAR402671 in Parkinson's disease patients carrying a glucocerebrosidase (GBA) gene mutation (MOVES-PD). US National Library of Medicine: ClinicalTrials.gov. https://clinicaltrials.gov/ct2/show/NCT02906020. Updated March 14, 2019. Accessed March 29, 2019.

57. Vision and mission. Parkinson's Foundation. www.parkinson.org/about -us/vision-and-mission. Accessed March 29, 2019.

58. Parkinson's Foundation announces new genetic initiative connecting Parkinson's genetic data with clinical care [press release]. Parkinson's Foundation. March 6, 2018. https://parkinson.org/about-us/Press-Room/Press -Releases/Parkinsons-Foundation-Announces-New-Genetic-Initiative -Connecting-Parkinsons-Genetic-Data-with-Clinical-Care. Accessed June 27, 2019.

59. Saunders-Pullman R, Mirelman A, Alcalay RN, et al. Progression in the LRRK2-associated Parkinson disease population. *JAMA Neurology.* 2018;75(3):312–319; Davis MY, Johnson CO, Leverenz JB, et al. Association of GBA mutations and the E326K polymorphism with motor and cognitive progression in Parkinson disease. *JAMA Neurology.* 2016;73(10):1217–1224; Genetics initiative. Parkinson's Foundation. www.parkinson.org/research /genetics-initiative. Accessed March 29, 2019.

60. Rinaldo CR, Jr. Passive immunization against poliomyelitis: the Hammon gamma globulin field trials, 1951–1953. *American Journal of Public Health*. 2005;95(5):790–799.

61. Ibid.

62. Jankovic J, Goodman I, Safirstein B, et al. Safety and tolerability of multiple ascending doses of PRX002/RG7935, an anti-α-synuclein monoclonal antibody, in patients with Parkinson disease: a randomized clinical trial. *JAMA Neurology*. 2018;75(10):1206–1214.

63. Fletcher K. Trials to treatments: vaccines for Parkinson's. Medium. November 19, 2018. https://medium.com/parkinsons-uk/vaccines-for-parkinsons-b09634c84017. Accessed May 20, 2019.

64. Jankovic J, Goodman I, Safirstein B, et al. Safety and tolerability of multiple ascending doses of PRX002/RG7935, an anti-α-synuclein monoclonal antibody, in patients with Parkinson disease: a randomized clinical trial. *JAMA Neurology*. 2018;75(10):1206–1214; Manfredsson FP, Tansey MG, Golde TE. Challenges in passive immunization strategies to treat Parkinson disease. *JAMA Neurology*. 2018;75(10):1180–1181.

65. AFFiRiS AG. AFFiRiS announces top line results of first-in-human clinical study using AFFITOPE PD03A, confirming immunogenicity and safety profile in Parkinson's disease patients. Cision PR Newswire. June 7, 2017. www.prnewswire.com/news-releases/affiris-announces-top-line-results-of-first-in-human-clinical-study-using-affitope-pd03a-confirming-immunogenicity-and-safety-profile-in-parkinsons-disease-patients-627025511.html. Accessed June 27, 2019; Affitope PD03A. Parkinsons's News Today. https://parkinsonsnewstoday.com/affitope-pd03a. Accessed March 29, 2019.

66. Kuhl MM. Vaccine for Parkinson's reports positive results from boost study. *FoxFeed Blog*. September 7, 2016. www.michaeljfox.org/news/vaccine-parkinsons-reports-positive-results-boost-study. Accessed June 27, 2019.

67. Dieleman JL, Baral R, Birger M, et al. US spending on personal health care and public health, 1996–2013. *JAMA*. 2016;316(24):2627–2646.

68. Olanow CW, Rascol O, Hauser R, et al. A double-blind, delayed-start trial of rasagiline in Parkinson's disease. *New England Journal of Medicine*. 2009;361(13):1268–1278; Parkinson Study Group. A controlled trial of rasagiline in early Parkinson disease: the TEMPO Study. *Archives of Neurology*. 2002;59(12):1937–1943.

69. Olanow CW, Rascol O, Hauser R, et al. A double-blind, delayed-start trial of rasagiline in Parkinson's disease. *New England Journal of Medicine*. 2009;361(13):1268–1278; Blandini F, Armentero MT, Fancellu R, Blaugrund E, Nappi G. Neuroprotective effect of rasagiline in a rodent model of Parkinson's disease. *Experimental Neurology*. 2004;187(2):455–459.

70. Blandini F, Armentero MT, Fancellu R, Blaugrund E, Nappi G. Neuroprotective effect of rasagiline in a rodent model of Parkinson's disease. *Experimental Neurology.* 2004;187(2):455–459.

71. 45 million Americans forego medications due to costs, new analysis shows—9 times the rate of the UK [press release]. Prescription Justice. February 6 2017. https://prescriptionjustice.org/press_release/45-million-americans-forego-medications-due-to-costs-new-analysis-shows-9-times-the-rate-of-the-uk. Accessed June 27, 2019.

72. DiMasi JA, Grabowski HG, Hansen RW. Innovation in the pharmaceutical industry: new estimates of R&D costs. *Journal of Health Economics.* 2016;47:20–33.

73. Adams CP, Brantner VV. Estimating the cost of new drug development: is it really $802 million? *Health Affairs.* 2006;25(2):420–428; Morgan S, Grootendorst P, Lexchin J, Cunningham C, Greyson D. The cost of drug development: a systematic review. *Health Policy.* 2011;100(1):4–17.

74. Adams CP, Brantner VV. Estimating the cost of new drug development: is it really $802 million? *Health Affairs.* 2006;25(2):420–428; Wong CH, Siah KW, Lo AW. Estimation of clinical trial success rates and related parameters. *Biostatistics.* 2018;20(2):273–286; Thomas DW, Burns J, Audette J, Carroll A, Dow-Hygelund C, Hay M. *Clinical Development Success Rates 2006–2015.* Biotechnology Innovation Organization. June 2016. www.bio.org/sites/default/files/Clinical%20Development%20Success%20Rates%202006-2015%20-%20BIO,%20Biomedtracker,%20Amplion%202016.pdf. Accessed June 28, 2019.

75. Kesselheim AS, Avorn J, Sarpatwari A. The high cost of prescription drugs in the United States: origins and prospects for reform. *JAMA.* 2016;316(8):858–871; Gaffney A, Lexchin J. Healing an ailing pharmaceutical system: prescription for reform for United States and Canada. *BMJ.* 2018;361:k1039.

76. Bunis D. Medicare "doughnut hole" will close in 2019. AARP. February 9, 2018. www.aarp.org/health/medicare-insurance/info-2018/part-d-donut-hole-closes-fd.html. Accessed March 29, 2019.

77. Pasternack A. This pioneering $475,000 cancer drug comes with a money-back guarantee. Fast Company. August 31, 2017. www.fastcompany.com/40461214/how-novartis-is-defending-the-record-475000-price-of-its-pioneering-gene-therapy-cancer-drug-car-t-kymriah. Accessed March 29, 2019.

78. Kesselheim AS, Avorn J, Sarpatwari A. The high cost of prescription drugs in the United States: origins and prospects for reform. *JAMA.* 2016;316(8):858–871.

79. Herper M. Solving the drug patent problem. *Forbes*. May 2, 2002. www
.forbes.com/2002/05/02/0502patents.html#3199af4317bc. Accessed June
28, 2019; Fox E. How pharma companies game the system to keep drugs
expensive. *Harvard Business Review*. April 6, 2017. https://hbr.org/2017
/04/how-pharma-companies-game-the-system-to-keep-drugs-expensive.
Accessed June 28, 2019.

80. Table 1, cost comparison table for drugs in early and advanced idiopathic
Parkinson disease. In: *Rotigotine (Neupro) (Transdermal Patch)* [Internet].
Canadian Agency for Drugs and Technologies in Health; 2016; Carbidopa/
Levodopa. GoodRx. www.goodrx.com/carbidopa-levodopa. Accessed March
29, 2019.

81. *Atlas: Country Resources for Neurological Disorders*. World Health Organization; 2004.

82. *WHO Model List of Essential Medicines*. World Health Organization;
April 2015.

83. Branswell H. Health officials set to release a list of drugs everyone
on earth should be able to access. STAT. June 5, 2017. www.statnews.com
/2017/06/05/essential-medicines-list-who. Accessed March 28, 2019.

84. Katzenschlager R, Evans A, Manson A, et al. *Mucuna pruriens* in Parkinson's disease: a double blind clinical and pharmacological study. *Journal
of Neurology, Neurosurgery & Psychiatry*. 2004;75(12):1672.

85. Ibid.; Cilia R, Laguna J, Cassani E, et al. *Mucuna pruriens* in Parkinson
disease: a double-blind, randomized, controlled, crossover study. *Neurology*.
2017;89(5):432–438.

CHAPTER 9: TAKING CHARGE

1. The Cure Parkinson's Trust and Parkinson's community mourn the loss
of their champion Tom Isaacs. Cure Parkinson's Trust. June 19, 2017. www
.cureparkinsons.org.uk/news/cpt-pd-community-mourns-the-loss. Accessed
March 14, 2019; Tom Isaacs. *The Times*. July 5, 2017; Obituary.

2. Palfreman J. CPT Co founder Tom Isaacs featured in the Journal of
Parkinson's Disease. Cure Parkinson's Trust. October 30, 2013. www.cure
parkinsons.org.uk/News/cpt-co-founder-tom-isaacs-featured-in-the-journal
-of-parkinsons-disease. Accessed March 27, 2019.

3. Our story. Cure Parkinson's Trust. www.cureparkinsons.org.uk/our-story.
Accessed March 27, 2019.

4. Palfreman J. CPT Co founder Tom Isaacs featured in the Journal of
Parkinson's Disease. Cure Parkinson's Trust. October 30, 2013. www

.cureparkinsons.org.uk/News/cpt-co-founder-tom-isaacs-featured-in-the -journal-of-parkinsons-disease. Accessed March 27, 2019.

5. The Cure Parkinson's Trust and Parkinson's community mourn the loss of their champion Tom Isaacs. Cure Parkinson's Trust. June 19, 2017. www .cureparkinsons.org.uk/news/cpt-pd-community-mourns-the-loss. Accessed March 14, 2019; Ethics. University of Notre Dame Center for Stem Cells and Regenerative Medicine. https://stemcell.nd.edu/ethics. Accessed March 27, 2019; No research justifies the use of human embryos, Pope Francis says. Catholic News Agency. May 18, 2017. www.catholicnewsagency.com/news /no-research-justifies-the-use-of-human-embryos-pope-francis-says-22203. Accessed March 27, 2019.

6. The Cure Parkinson's Trust and Parkinson's community mourn the loss of their champion Tom Isaacs. Cure Parkinson's Trust. June 19, 2017. www .cureparkinsons.org.uk/news/cpt-pd-community-mourns-the-loss. Accessed March 14, 2019.

7. Ibid.; Tom Isaacs memorial. World Parkinson Coalition. www.world pdcoalition.org/page/TomIsaacs. Accessed March 27, 2019.

8. Grisales C. Veterans, families demand that EPA ban toxic chemical found in tap water at US military bases. Stars and Stripes. August 1, 2018. www.stripes .com/news/veterans-families-demand-that-epa-ban-toxic-chemical-found -in-tap-water-at-us-military-bases-1.540501. Accessed February 11, 2019.

9. Ibid.

10. Returning EPA to its core mission. US Environmental Protection Agency. www.epa.gov/home/returning-epa-its-core-mission. Updated July 6, 2018. Accessed March 27, 2019.

11. Falce A. Congress increases research funding, shows support for care partners. *FoxFeed Blog*. September 28, 2018. www.michaeljfox.org/news /congress-increases-research-funding-shows-support-care-partners. Accessed June 28, 2019.

12. Intel co-founder Andrew S. Grove dedicates portion of his estate, up to $40 million, to Michael J. Fox Foundation [press release]. The Michael J. Fox Foundation. January 10, 2008. www.michaeljfox.org/publication/intel -co-founder-andrew-s-grove-dedicates-portion-his-estate-40-million-michael -j-fox. Accessed June 28, 2019.

13. Reuters. Intel pioneer and Holocaust survivor Andy Grove dies at 79. Haaretz. March 22, 2016. www.haaretz.com/jewish/intel-pioneer-and -holocaust-survivor-andy-grove-dies-at-79-1.5421022. Accessed March 27, 2019; Newton C. Legendary former Intel CEO Andy Grove is dead at 79. The Verge. March 21, 2016. www.theverge.com/2016/3/21/11280004/andy -grove-intel-ceo-dies. Accessed March 27, 2019.

14. Andrew Grove. Wikipedia. https://en.wikipedia.org/wiki/Andrew_ Grove. Updated February 10, 2019. Accessed February 11, 2019.

15. Dolan KA. Andy Grove's last stand. *Forbes*. January 11, 2008. www .forbes.com/forbes/2008/0128/070.html#462dbe0b641e. Accessed June 28, 2019.

16. Ibid.

17. Pollack A, Levy J. The New York Times: Andy Grove's prescription for health care. *FoxFeed Blog*. November 18, 2009. www.michaeljfox.org/news/new -york-times-andy-groves-prescription-health-care. Accessed June 28, 2019.

18. Olavsrud T. Intel turns to wearables, big data to fight Parkinson's. CIO .com. October 28, 2014. www.cio.com/article/2839555/intel-turns-to -wearables-big-data-to-fight-parkinsons.html. Accessed March 28, 2019; Un- gerleider N. Why Intel and The Michael J. Fox Foundation are teaming up to create wearable tech for Parkinson's. Fast Company. August 15, 2014. www .fastcompany.com/3034433/why-intel-and-the-michael-j-fox-foundation -are-teaming-up-to-create-wearable-tech-for-parkin. Accessed March 28, 2019.

19. Andy Grove: we remember six key facts about the former Intel CEO and visionary. NexChange. 2016. https://nexchange.com/article/9009. Ac- cessed March 28, 2019.

20. McGregor J. This moving tribute explains why former Intel CEO Andy Grove was revered by so many. *Washington Post*. March 22, 2016. www .washingtonpost.com/news/on-leadership/wp/2016/03/22/this -moving-tribute-explains-why-former-intel-ceo-andy-grove-was-so-revered -by-so-many/?noredirect=on&utm_term=.3ba0dee8f248. Accessed June 28, 2019.

21. Goetz CG, Stebbins GT, Wolff D, et al. Testing objective measures of motor impairment in early Parkinson's disease: feasibility study of an at-home testing device. *Movement Disorders*. 2009;24(4):551–556.

22. Shinal J. Andy Grove remembered as "bridge builder." *USA Today*. June 20, 2016.

23. Michael J. Fox Foundation for Parkinson's Research. Tuck School of Business at Dartmouth. http://mytuck.dartmouth.edu/show_module_fw2 .aspx?sid=1353&gid=5&control_id=11858&nologo=1&cvprint=1&page _id=4475&crid=0&viewas=user. Accessed March 28, 2019.

24. Teichholtz H. The Michael J. Fox Foundation mourns passing of Andy Grove. *FoxFeed Blog*. March 22, 2016. www.michaeljfox.org/news/michael-j -fox-foundation-mourns-passing-andy-grove. Accessed June 28, 2019.

25. McGregor J. This moving tribute explains why former Intel CEO Andy Grove was revered by so many. *Washington Post*. March 22, 2016. www

.washingtonpost.com/news/on-leadership/wp/2016/03/22/this-moving
-tribute-explains-why-former-intel-ceo-andy-grove-was-so-revered-by
-so-many/?noredirect=on&utm_term=.3ba0dee8f248. Accessed June 28,
2019; Program information. Andy Grove Scholarship for Intel Employees'
Children. www.scholarsapply.org/intel/instructions.php. Accessed March 28,
2019; Burgelman R, Meza P. Grove Scholars Program: putting rungs back on
the ladder. Stanford Graduate School of Business. 2005. www.gsb.stanford
.edu/faculty-research/case-studies/grove-scholars-program-putting-rungs
-back-ladder. Accessed March 28, 2019; City College mourns distinguished
alumnus Andrew Grove '60. City College of New York. March 22, 2016.
www.ccny.cuny.edu/news/city-college-mourns-distinguished-alumnus
-andrew-grove-%E2%80%9960. Accessed March 28, 2019.

26. Kaiser J. Final 2018 budget bill eases biomedical researchers' policy
worries. *Science*. March 22, 2018. American Association for the Advancement
of Science. www.sciencemag.org/news/2018/03/final-2018-budget-bill
-eases-biomedical-researchers-policy-worries. Accessed June 27, 2019.

27. Estimates of funding for various research, condition, and disease categories (RCDC). National Institutes of Health. 2018. https://report.nih
.gov/categorical_spending.aspx. Updated May 18, 2018. Accessed March 31,
2019; Biomedical Research and Development Price Index. Bureau of Economic Analysis. 2018. https://officeofbudget.od.nih.gov/gbipriceindexes.html.
Updated January 2018. Accessed October 15, 2018.

28. Marras C, Beck JC, Bower JH, et al. Prevalence of Parkinson's disease
across North America. *npj Parkinson's Disease*. 2018;4(1):21.

29. Estimates of funding for various research, condition, and disease categories (RCDC). National Institutes of Health. 2018. https://report.nih
.gov/categorical_spending.aspx. Updated May 18, 2018. Accessed March 31,
2019; Biomedical Research and Development Price Index. Bureau of Economic Analysis. 2018. https://officeofbudget.od.nih.gov/gbipriceindexes
.html. Updated January 2018. Accessed October 15, 2018; Marras C, Beck
JC, Bower JH, et al. Prevalence of Parkinson's disease across North America.
npj Parkinson's Disease. 2018;4(1):21.

30. Chun D. First sign of Parkinson's hit Fox during filming. Gainesville
Sun. May 15, 2009. www.gainesville.com/news/20090515/first-sign-of
-parkinsons-hit-fox-during-filming. Accessed March 28, 2019.

31. Raffalli M. Michael J. Fox: working towards cure for Parkinson's "one of
the great gifts of my life." CBS News. July 8, 2018. www.cbsnews.com/news
/michael-j-fox-working-towards-cure-for-parkinsons-one-of-the-great-gifts-
of-my-life. Accessed March 28, 2019.

32. IRS Form 990. The Breast Cancer Research Foundation Inc. 2016. https://projects.propublica.org/nonprofits/display_990/133727250/2017_03_EO%2F13-3727250_990_201606; IRS Form 990. Alzheimer's Disease & Related Disorders Association, Inc. 2016. www.alz.org/media/Documents/form-990-fy-2017.pdf; IRS Form 990. JDRF International. 2016. https://1x5o5mujiug388ttap1p8s17-wpengine.netdna-ssl.com/wp-content/uploads/2018/01/JDRF-FY2017-990-Tax-Return.pdf; IRS Form 990, Arthritis Foundation Inc. 2016. www.arthritis.org/Documents/Sections/About-Us/2016-National-Office-990.pdf; IRS Form 990. The Michael J. Fox Foundation for Parkinson's Research. 2016. https://files.michaeljfox.org/2016-990-MJFF.pdf.

33. Nocera J. The Michael J. Fox Foundation gets results: the latest treatment for Parkinson's disease was made possible by the actor's audacious mission. Bloomberg. January 23, 2019. www.bloomberg.com/opinion/articles/2019-01-23/parkinson-s-inhaler-the-michael-j-fox-foundation-gets-results. Accessed February 11, 2019.

34. Patel AB, Jimenez-Shahed J. Profile of inhaled levodopa and its potential in the treatment of Parkinson's disease: evidence to date. *Neuropsychiatric Disease and Treatment.* 2018;14:2955–2964.

35. King S. *Pink Ribbons, Inc.: Breast Cancer and the Politics of Philanthropy.* 1st ed. University of Minnesota Press; 2008.

36. Tanner CM, Kamel F, Ross GW, et al. Rotenone, paraquat, and Parkinson's disease. *Environmental Health Perspectives.* 2011;119(6):866–872; Goldman SM, Quinlan PJ, Ross GW, et al. Solvent exposures and Parkinson disease risk in twins. *Annals of Neurology.* 2012;71(6):776–784; Tanner CM, Ottman R, Goldman SM, et al. Parkinson disease in twins: an etiologic study. *JAMA.* 1999;281(4):341–346.

37. Kuhl MM. New $6 million program looks at causes of Parkinson's. *FoxFeed Blog.* March 22, 2018. www.michaeljfox.org/news/new-6-million-program-looks-causes-parkinsons. Accessed June 28, 2019.

38. State Cancer Profiles. https://statecancerprofiles.cancer.gov. Accessed March 28, 2019.

39. Strickland D, Bertoni JM. Parkinson's prevalence estimated by a state registry. Movement Disorders. 2004;19(3):318–323; Wright Willis A, Evanoff BA, Lian M, Criswell SR, Racette BA. Geographic and ethnic variation in Parkinson disease: a population-based study of US Medicare beneficiaries. Neuroepidemiology. 2010;34(3):143–151.

40. Parkinson Disease Registry. Utah Department of Health. 2018. www.updr.org. Accessed January 4, 2019; Parkinson's Disease Registry. Nebraska Department of Health & Human Services. 2015. http://dhhs.ne.gov/public

health/Pages/ced_parkinsons_index.aspx. Updated February 7, 2018. Accessed January 4, 2019.

41. Tanner CM. California's Parkinson's Disease Registry pilot project—coordination center and Northern California ascertainment. Defense Technical Information Center. March 2014. https://apps.dtic.mil/dtic/tr/fulltext/u2/a601866.pdf. Accessed June 28, 2019.

42. California Parkinson's Disease Registry. California Department of Public Health. Chronic Disease Surveillance and Research Branch (CDSRB). www.cdph.ca.gov/Programs/CCDPHP/DCDIC/CDSRB/Pages/California-Parkinson%27s-Disease-Registry.aspx. Updated October 1, 2018. Accessed October 15, 2018.

43. Parkinson Disease Registry. Utah Department of Health. 2018. www.updr.org. Accessed January 4, 2019; Parkinson's Disease Registry. Nebraska Department of Health & Human Services. 2015. http://dhhs.ne.gov/publichealth/Pages/ced_parkinsons_index.aspx. Updated February 7, 2018. Accessed January 4, 2019.

44. Strickland D, Bertoni JM. Parkinson's prevalence estimated by a state registry. *Movement Disorders.* 2004;19(3):318–323.

45. GBD 2016 Parkinson's Disease Collaborators. Global, regional, and national burden of Parkinson's disease, 1990–2016: a systematic analysis for the Global Burden of Disease Study 2016. *Lancet Neurology.* 2018;17(11):939–953.

46. Singleton AB, Farrer M, Johnson J, et al. α-synuclein locus triplication causes Parkinson's disease. *Science.* 2003;302(5646):841–841.

47. Paisán-Ruíz C, Jain S, Evans EW, et al. Cloning of the gene containing mutations that cause PARK8-linked Parkinson's disease. *Neuron.* 2004;44(4):595–600.

48. Farrer MJ. Genetics of Parkinson disease: paradigm shifts and future prospects. *Nature Reviews Genetics.* 2006;7(4):306–318; Simón-Sánchez J, Schulte C, Bras JM, et al. Genome-wide association study reveals genetic risk underlying Parkinson's disease. *Nature Genetics.* 2009;41:1308; Dawson TM, Ko HS, Dawson VL. Genetic animal models of Parkinson's disease. *Neuron.* 2010;66(5):646–661; Singleton AB, Hardy JA, Gasser T. The birth of the modern era of Parkinson's disease genetics. *Journal of Parkinson's Disease.* 2017;7(Suppl 1):S87–S93.

49. Understanding targeted therapy. American Society of Clinical Oncology. Cancer.net. January 2019. www.cancer.net/navigating-cancer-care/how-cancer-treated/personalized-and-targeted-therapies/understanding-targeted-therapy. Accessed March 28, 2019.

50. Jackson SE, Chester JD. Personalised cancer medicine. *International Journal of Cancer.* 2015;137(2):262–266.

51. Singleton AB, Hardy JA, Gasser T. The birth of the modern era of Parkinson's disease genetics. *Journal of Parkinson's Disease.* 2017;7(Suppl 1):S87–S93.

52. Goker-Alpan O, Schiffmann R, LaMarca ME, Nussbaum RL, McInerney-Leo A, Sidransky E. Parkinsonism among Gaucher disease carriers. *Journal of Medical Genetics.* 2004;41(12):937–940.

53. Study goals. Parkinson's Progression Markers Initiative. www.ppmi -info.org/about-ppmi/study-goals. Accessed March 28, 2019.

54. Access Data & Specimens. Parkinson's Progression Markers Initiative. www.ppmi-info.org/access-data-specimens. Accessed February 11, 2019.

55. PPMI Clinical Study. The Michael J. Fox Foundation for Parkinson's Research. www.michaeljfox.org/ppmi-clinical-study. Accessed October 15, 2018.

56. Gash DM, Rutland K, Hudson NL, et al. Trichloroethylene: parkinsonism and complex 1 mitochondrial neurotoxicity. *Annals of Neurology.* 2008;63(2):184–192.

57. Ibid.

58. Haddad D, Nakamura K. Understanding the susceptibility of dopamine neurons to mitochondrial stressors in Parkinson's disease. *FEBS Letters.* 2015;589(24 Pt A):3702–3713.

59. Xu L, Pu J. Alpha-synuclein in Parkinson's disease: from pathogenetic dysfunction to potential clinical application. *Parkinson's Disease.* 2016;2016:1720621.

60. Bendor J, Logan T, Edwards RH. The function of α-synuclein. *Neuron.* 2013;79(6):10.1016/j.neuron.2013.1009.1004.

61. Smith WW, Pei Z, Jiang H, Dawson VL, Dawson TM, Ross CA. Kinase activity of mutant LRRK2 mediates neuronal toxicity. *Nature Neuroscience.* 2006;9(10):1231–1233; Martin I, Kim JW, Dawson VL, Dawson TM. LRRK2 pathobiology in Parkinson's disease. *Journal of Neurochemistry.* 2014;131(5):554–565.

62. Martin I, Kim JW, Dawson VL, Dawson TM. LRRK2 pathobiology in Parkinson's disease. *Journal of Neurochemistry.* 2014;131(5):554–565.

63. Moses H, Dorsey ER, Matheson DM, Thier SO. Financial anatomy of biomedical research. *JAMA.* 2005;294(11):1333–1342; Moses H, Matheson DM, Cairns-Smith S, George BP, Palisch C, Dorsey E. The anatomy of medical research: US and international comparisons. *JAMA.* 2015;313(2):174–189.

64. Moses H, Dorsey ER, Matheson DM, Thier SO. Financial anatomy of biomedical research. *JAMA*. 2005;294(11):1333–1342.

65. 2018 PhRMA Annual Membership Survey. Pharmaceutical Research and Manufacturers of America. July 26, 2018. www.phrma.org/report /2018-phrma-annual-membership-survey. Accessed June 28, 2019.

66. Moses H, Dorsey ER, Matheson DM, Thier SO. Financial anatomy of biomedical research. *JAMA*. 2005;294(11):1333–1342; Moses H, Matheson DM, Cairns-Smith S, George BP, Palisch C, Dorsey E. The anatomy of medical research: US and international comparisons. *JAMA*. 2015;313(2):174–189.

67. Moses H, Dorsey ER, Matheson DM, Thier SO. Financial anatomy of biomedical research. *JAMA*. 2005;294(11):1333–1342.

68. Cannon JR, Greenamyre JT. Gene-environment interactions in Parkinson's disease: specific evidence in humans and mammalian models. *Neurobiology of Disease*. 2013;57:38–46.

69. Ibid.

70. Ibid.

71. Di Maio R, Hoffman EK, Rocha EM, et al. LRRK2 activation in idiopathic Parkinson's disease. *Science Translational Medicine*. 2018;10(451); Cannon JR, Greenamyre JT. Gene-environment interactions in Parkinson's disease: specific evidence in humans and mammalian models. *Neurobiology of Disease*. 2013;57:38–46; Lee J-W, Cannon JR. LRRK2 mutations and neurotoxicant susceptibility. *Experimental Biology and Medicine*. 2015;240(6):752–759.

72. Hughes AJ, Daniel SE, Kilford L, Lees AJ. Accuracy of clinical diagnosis of idiopathic Parkinson's disease: a clinico-pathological study of 100 cases. *Journal of Neurology, Neurosurgery, and Psychiatry*. 1992;55(3):181–184; Rizzo G, Copetti M, Arcuti S, Martino D, Fontana A, Logroscino G. Accuracy of clinical diagnosis of Parkinson disease: a systematic review and meta-analysis. *Neurology*. 2016;86(6):566–576.

73. Hughes AJ, Daniel SE, Kilford L, Lees AJ. Accuracy of clinical diagnosis of idiopathic Parkinson's disease: a clinico-pathological study of 100 cases. *Journal of Neurology, Neurosurgery, and Psychiatry*. 1992;55(3):181–184; Schrag A, Ben-Shlomo Y, Quinn N. How valid is the clinical diagnosis of Parkinson's disease in the community? *Journal of Neurology, Neurosurgery & Psychiatry*. 2002;73(5):529.

74. Rizzo G, Copetti M, Arcuti S, Martino D, Fontana A, Logroscino G. Accuracy of clinical diagnosis of Parkinson disease: a systematic review and meta-analysis. *Neurology*. 2016;86(6):566–576.

75. MDS-UPDRS. International Parkinson and Movement Disorder Society. July 1, 2008. www.movementdisorders.org/MDS-Files1/PDFs/Rating -Scales/MDS-UPDRS_English_FINAL_Updated_June2019.pdf.

76. Dolan KA. Andy Grove's last stand. *Forbes.* January 11, 2008. www .forbes.com/forbes/2008/0128/070.html#462dbe0b641e. Accessed June 28, 2019.

77. Little MA, McSharry PE, Roberts SJ, Costello DA, Moroz IM. Exploiting nonlinear recurrence and fractal scaling properties for voice disorder detection. *BioMedical Engineering OnLine.* 2007;6(1):23.

78. Little MA, McSharry PE, Hunter EJ, Spielman J, Ramig LO. Suitability of dysphonia measurements for telemonitoring of Parkinson's disease. *IEEE Transactions on Biomedical Engineering.* 2009;56(4):1015–1022.

79. Tsanas A, Little MA, McSharry PE, Spielman J, Ramig LO. Novel speech signal processing algorithms for high-accuracy classification of Parkinson's disease. *IEEE Transactions on Biomedical Engineering.* 2012;59(5):1264–1271.

80. Little MA, McSharry PE, Hunter EJ, Spielman J, Ramig LO. Suitability of dysphonia measurements for telemonitoring of Parkinson's disease. *IEEE Transactions on Biomedical Engineering.* 2009;56(4):1015–1022; Tsanas A, Little MA, McSharry PE, Spielman J, Ramig LO. Novel speech signal processing algorithms for high-accuracy classification of Parkinson's disease. *IEEE Transactions on Biomedical Engineering.* 2012;59(5):1264–1271; Tsanas A, Little MA, McSharry PE, Ramig LO. Accurate telemonitoring of Parkinson's disease progression by noninvasive speech tests. *IEEE Transactions on Biomedical Engineering.* 2010;57(4):884–893; Tsanas A, Little MA, McSharry PE, Ramig LO. Nonlinear speech analysis algorithms mapped to a standard metric achieve clinically useful quantification of average Parkinson's disease symptom severity. *Journal of the Royal Society Interface.* 2011;8(59):842–855.

81. Little MA. A test for Parkinson's with a phone call [video file]. TEDGlobal 2012. June 2012. www.ted.com/talks/max_little_a_test_for _parkinson_s_with_a_phone_call. Accessed June 28, 2019.

82. Arora S, Venkataraman V, Zhan A, et al. Detecting and monitoring the symptoms of Parkinson's disease using smartphones: a pilot study. *Parkinsonism & Related Disorders.* 2015;21(6):650–653.

83. Ibid.

84. Zhan A, Mohan S, Tarolli C, et al. Using smartphones and machine learning to quantify Parkinson disease severity: the mobile Parkinson disease score. *JAMA Neurology.* 2018;75(7):876–880.

85. Bot BM, Suver C, Neto EC, et al. The mPower study, Parkinson disease mobile data collected using ResearchKit. *Scientific Data.* 2016;3:160011.

86. McLeonida. ResearchKit demo by Jeff Williams at Apple special event, March 2015 [video file]. YouTube. March 11, 2015. www.youtube.com/watch?v=O0gcEFjQNGk. Accessed June 28, 2019.

87. Dorsey ER, Beck CA, Adams M, et al. Communicating clinical trial results to research participants. *Archives of Neurology.* 2008;65(12):1590–1595.

88. Writing Group for the NINDS Exploratory Trials in Parkinson Disease Investigators. Effect of creatine monohydrate on clinical progression in patients with Parkinson disease: a randomized clinical trial. *JAMA.* 2015;313(6):584–593.

89. Dorsey ER, Chan Y-FY, McConnell MV, Shaw SY, Trister AD, Friend SH. The use of smartphones for health research. *Academic Medicine.* 2017;92(2):157–160.

90. Turakhia MP, Desai M, Hedlin H, et al. Rationale and design of a large-scale, app-based study to identify cardiac arrhythmias using a smartwatch: the Apple Heart Study. *American Heart Journal.* 2019;207:66–75.

91. Farr C. The Apple Watch just got a lot better at tracking symptoms of Parkinson's disease. CNBC. June 9, 2018. www.cnbc.com/2018/06/09/apple-watch-adds-tech-to-track-parkinsons-disease.html. Updated June 10, 2018. Accessed November 10, 2018.

92. Collaboration with Verily aims to deepen Parkinson's understanding through digital health tools. *FoxFeed Blog.* May 9, 2018. www.michaeljfox.org/news/collaboration-verily-aims-deepen-parkinsons-understanding-through-digital-health-tools?collaboration-with-verily-aims-to-deepen-parkinson-understanding-through-digital-health-tools=. Accessed June 28, 2019; About the study. Personalized Parkinson Project. www.parkinson opmaat.nl/studie. Accessed May 20, 2019.

93. Shults CW, Oakes D, Kieburtz K, et al. Effects of coenzyme q10 in early Parkinson disease: evidence of slowing of the functional decline. *Archives of Neurology.* 2002;59(10):1541–1550; Parkinson Study Group QE3 Investigators. A randomized clinical trial of high-dosage coenzyme Q10 in early Parkinson disease: no evidence of benefit. *JAMA Neurology.* 2014;71(5):543–552.

94. Adams CP, Brantner VV. Estimating the cost of new drug development: is it really $802 million? *Health Affairs.* 2006;25(2):420–428; Dorsey ER, Papapetropoulos S, Xiong M, Kieburtz K. The first frontier: digital biomarkers for neurodegenerative disorders. *Digital Biomarkers.* 2017;1(1):6–13; Dorsey ER, Venuto C, Venkataraman V, Harris DA, Kieburtz K. Novel methods and technologies for 21st-century clinical trials: a review. *JAMA*

Neurology. 2015;72(5):582–588; CNS drugs take longer to develop and have lower success rates than other drugs [press release]. Tufts Center for the Study of Drug Development. November 4, 2014. https://static1.squarespace .com/static/5a9eb0c8e2ccd1158288d8dc/t/5aa2bf604192023932fe1561 /1520615264660/PR-NOVDEC14.pdf. Accessed June 28, 2019.

95. Dwyer C. Pfizer halts research into Alzheimer's and Parkinson's treatments. *The Two-Way.* January 8, 2018. www.npr.org/sections/thetwo -way/2018/01/08/576443442/pfizer-halts-research-efforts-into-alzheimers -and parkinsons-treatments. Accessed June 28, 2019.

96. Ibid.

97. Taylor P. Backed by Bain, Pfizer loads prime CNS assets into new biotech. FierceBiotech. October 23, 2018. www.fiercebiotech.com/biotech /embargo-8am-edt-backed-by-bain-pfizer-loads-prime-cns-assets-into-new -biotech. Accessed February 11, 2019.

98. Knabe K. Pfizer's Parkinson's drugs will be developed by start-up Cerevel Therapeutics. *FoxFeed Blog.* October 24, 2018. www.michaeljfox.org /news/pfizers-parkinsons-drugs-will-be-developed-start-cerevel-therapeutics ?fbclid=IwAR3Wb4f_I6J4AqORHZCpJZeYCzGAdymx1jSxWVfu ZCKTq_-UEWAvY_bHGhI&pfizer-parkinson-drugs-will-be-developed -by-start-up-cerevel-therapeutics=. Accessed June 28, 2019.

99. Lipsmeier F, Taylor KI, Kilchenmann T, et al. Evaluation of smart-phone-based testing to generate exploratory outcome measures in a phase 1 Parkinson's disease clinical trial. *Movement Disorders.* 2018;33(8):1287–1297.

100. 221AD301 phase 3 study of aducanumab (BIIB037) in early Alz-heimer's disease (ENGAGE). US National Library of Medicine. Clinical Trials.gov. June 23, 2015. https://clinicaltrials.gov/ct2/show/study/NCT 02477800. Updated August 6, 2018. Accessed March 28, 2019; Henriques C. MJFF partners with Verily to use data collection watch in PPMI study to advance Parkinson's research. Parkinson's News Today. May 14, 2018. https://parkinsonsnewstoday.com/2018/05/14/mjff-verily-collaborate -advance-parkinsons-research-using-study-watch-ppmi. Accessed March 28, 2019.

101. Little MA. A test for Parkinson's with a phone call [video file]. TEDGlobal 2012. June 2012. www.ted.com/talks/max_little_a_test_for _parkinson_s_with_a_phone_call. Accessed June 28, 2019.

102. den Brok MGHE, van Dalen JW, van Gool WA, Moll van Charante EP, de Bie RMA, Richard E. Apathy in Parkinson's disease: a systematic re-view and meta-analysis. *Movement Disorders.* 2015;30(6):759–769; Patient

advocates respond to the PD pandemic. *WPC Blog*. December 19, 2017. www.worldpdcongress.org/home/2017/12/15/patient-advocates-respond -to-the-pd-pandemic. Accessed June 28, 2019.

103. Gordon L. With Words as My Oars: Paddling the Parkinson's Rapids. www.leonoregordonpdspeak.com. Accessed February 11, 2019; Wheatley Alumni Newsletter: Number 25: November 25, 2018. Wheatley School Alumni Association. November 25, 2018. www.wheatleyalumni.org /BlogPost/Blogpost-20181125-25.html. Accessed June 28, 2019.

104. Season 2: Mental Health. ParkinsonTV. July 2018. https:// parkinsontv.org/two. Accessed February 11, 2019.

105. What are the odds of dying from. National Safety Council. 2017. www.nsc.org/work-safety/tools-resources/injury-facts/chart. Accessed February 11, 2019.

106. Driver JA, Logroscino G, Gaziano JM, Kurth T. Incidence and remaining lifetime risk of Parkinson disease in advanced age. *Neurology*. 2009;72(5):432–438.

107. Mark Morris Dance Group. In memory of Leonore Gordon. Dance for PD. 2017. https://danceforparkinsons.org/photo-credits/permissions/in -memory-of-leonore-gordon. Accessed February 11, 2019.

CHAPTER 10: WITHIN OUR REACH

1. Bates C, Rowell A. *Tobacco Explained: The Truth About the Tobacco Industry*. Action on Smoking and Health; 1998.

2. Paraquat dichloride. US Environmental Protection Agency. www.epa .gov/ingredients-used-pesticide-products/paraquat-dichloride. Updated May 15, 2019. Accessed May 19, 2019; Paraquat Dichloride: One Sip Can Kill. US Environmental Protection Agency. Pesticide Worker Safety. https:// www.epa.gov/pesticide-worker-safety/paraquat-dichloride-one-sip-can-kill. Last updated May 8, 2019. Accessed March 25, 2019.

3. Tanner CM, Kamel F, Ross GW, et al. Rotenone, paraquat, and Parkinson's disease. *Environmental Health Perspectives*. 2011;119(6):866–872; Wang A, Costello S, Cockburn M, Zhang X, Bronstein J, Ritz B. Parkinson's disease risk from ambient exposure to pesticides. *European Journal of Epidemiology*. 2011;26(7):547–555; Costello S, Cockburn M, Bronstein J, Zhang X, Ritz B. Parkinson's disease and residential exposure to maneb and paraquat from agricultural applications in the Central Valley of California. *American Journal of Epidemiology*. 2009;169(8):919–926; Mc-

Cormack AL, Thiruchelvam M, Manning-Bog AB, et al. Environmental risk factors and Parkinson's disease: selective degeneration of nigral dopaminergic neurons caused by the herbicide paraquat. *Neurobiology of Disease.* 2002;10(2):119–127.

4. Pesticide National Synthesis Project: estimated annual agricultural pesticide use. US Geological Survey. https://water.usgs.gov/nawqa/pnsp/usage/maps/show_map.php?year=2015&map=PARAQUAT&hilo=H.Updated September 11, 2018. Accessed March 25, 2019.

5. Federal Insecticide, Fungicide, and Rodenticide Act (FIFRA) and Federal Facilities. US Environmental Protection Agency. www.epa.gov/enforcement/federal-insecticide-fungicide-and-rodenticide-act-fifra-and-federal-facilities. Updated January 29, 2018. Accessed March 29, 2019.

6. H.R.3817—To cancel the registration of all uses of the pesticide paraquat, and for other purposes. Congress.gov. www.congress.gov/bill/116th-congress/house-bill/3817/committees. Accessed July 19, 2019.

7. Freire C, Koifman S. Pesticide exposure and Parkinson's disease: epidemiological evidence of association. *NeuroToxicology.* 2012;33(5):947–971; Gatto NM, Cockburn M, Bronstein J, Manthripragada AD, Ritz B. Well-water consumption and Parkinson's disease in rural California. *Environmental Health Perspectives.* 2009;117(12):1912–1918; Chen T, Tan J, Wan Z, et al. Effects of commonly used pesticides in China on the mitochondria and ubiquitin-proteasome system in Parkinson's disease. *International Journal of Molecular Sciences.* 2017;18(12):2507; Pallotta MM, Ronca R, Carotenuto R, et al. Specific effects of chronic dietary exposure to chlorpyrifos on brain gene expression—a mouse study. *International Journal of Molecular Sciences.* 2017;18(11):2467; Rauh VA. Polluting developing brains—EPA failure on chlorpyrifos. *New England Journal of Medicine.* 2018;378(13):1171–1174.

8. Pesticide tolerances: chlorpyrifos. Regulations.gov. www.regulations.gov/document?D=EPA-HQ-OPP-2007-1005-0100. Accessed March 29, 2019.

9. Letters in support of September 2018 Department of Justice action on chlorpyrifos. US Environmental Protection Agency. www.epa.gov/ingredients-used-pesticide-products/letters-support-september-2018-department-justice-action. Updated September 24, 2018. Accessed March 29, 2019.

10. Ibid.

11. Pesticide tolerances: chlorpyrifos. Regulations.gov. www.regulations.gov/document?D=EPA-HQ-OPP-2007-1005-0100. Accessed March 29, 2019; Chlorpyrifos. US Environmental Protection Agency. www.epa.gov

/ingredients-used-pesticide-products/chlorpyrifos. Updated September 24, 2018. Accessed March 29, 2019; O'Neill E. This pesticide poisons kids, but it's still sprayed on Washington orchards. Oregon Public Broadcasting. December 16, 2018. www.opb.org/news/article/washington-orchards -northwest-pesticides-chlorpyrifos. Accessed March 29, 2019.

12. Pesticide information profile: chlorpyrifos. Extension Toxicology Network. September 1993. http://pmep.cce.cornell.edu/profiles/extoxnet /carbaryl-dicrotophos/chlorpyrifos-ext.html. Accessed March 30, 2019.

13. Parrish ML, Gardner RE. Is living downwind of a golf course a risk factor for parkinsonism? *Annals of Neurology*. 2012;72(6):984–984.

14. Ibid.

15. Wallace G. EPA is ordered to ban farm pesticide chlorpyrifos. CNN. August 9, 2018. www.cnn.com/2018/08/09/politics/epa-court-order-ban -pesticide-chlorpyrifos/index.html. Accessed March 30, 2019.

16. Ibid.

17. Chlorpyrifos. US Environmental Protection Agency. www.epa.gov /ingredients-used-pesticide-products/chlorpyrifos. Updated September 24, 2018. Accessed March 29, 2019; Cama T. Trump admin appeals ruling ordering EPA to ban pesticide. The Hill. September 24, 2018. https://thehill.com /policy/energy-environment/408173-trump-admin-appeals-ruling-ordering -epa-to-ban-pesticide. Accessed March 30, 2019.

18. California to ban controversial pesticide blamed for harming child brain development. CBS News. May 8, 2019. www.cbsnews.com/news/california -bans-chlorpyrifos-pesticide-agriculture-state-child-brain-development. Accessed May 20, 2019.

19. Friedman L. E.P.A. won't ban chlorpyrifos, pesticide tied to children's health problems. New York Times. July 18, 2019. www.nytimes.com /2019/07/18/climate/epa-chlorpyrifos-pesticide-ban.html?smid=nytcore-ios -share. Accessed July 19, 2019.

20. www.congress.gov/bill/116th-congress/house-bill/230.

21. Rauh VA, Garfinkel R, Perera FP, et al. Impact of prenatal chlorpyrifos exposure on neurodevelopment in the first 3 years of life among inner-city children. *Pediatrics*. 2006;118(6):e1845–e1859.

22. Ibid.

23. Rauh VA. Polluting developing brains—EPA failure on chlorpyrifos. *New England Journal of Medicine*. 2018;378(13):1171–1174.

24. Hertz-Picciotto I, Sass JB, Engel S, et al. Organophosphate exposures during pregnancy and child neurodevelopment: recommendations for essential policy reforms. *PLOS Medicine*. 2018;15(10):e1002671.

25. Roser M, Ritchie H. Fertilizer and pesticides. Our World in Data. https://ourworldindata.org/fertilizer-and-pesticides. Accessed March 25, 2019.

26. McCord CP. Toxicity of trichloroethylene. *Journal of the American Medical Association.* 1932;99(5):409.

27. Risk management for trichloroethylene (TCE). US Environmental Protection Agency. www.epa.gov/assessing-and-managing-chemicals-under-tsca/risk-management-trichloroethylene-tce. Updated December 14, 2017. Accessed November 16, 2018.

20. Case studies on safer alternatives for solvent degreasing applications. US Environmental Protection Agency. www.epa.gov/p2/case-studies-safer-alternatives-solvent-degreasing-applications. Updated November 28, 2017. Accessed March 30, 2019.

29. Kaplan S. In reversal, chemicals are cleared for use. *New York Times.* December 20, 2017:A16.

30. Bjorhus J. Minnesota considers banning TCE following emissions problem in White Bear Township. *Star Tribune.* March 15, 2019.

31. Water gremlin: trichloroethylene (TCE) area of concern. Minnesota Pollution Control Agency. www.pca.state.mn.us/air/water-gremlin-trichloroethylene-tce-area-concern. Accessed March 30, 2019.

32. Ibid.

33. Bjorhus J. Minnesota considers banning TCE following emissions problem in White Bear Township. *Star Tribune.* March 15, 2019; Reilly M. Legislature may ban widely used solvent TCE after chemical leak at Minnesota manufacturer. *Minneapolis/St. Paul Business Journal.* March 15, 2019. www.bizjournals.com/twincities/news/2019/03/15/legislature-may-ban-widely-used-solvent-tce-after.html. Accessed June 28, 2019.

34. Kinney DK, Barch DH, Chayka B, Napoleon S, Munir KM. Environmental risk factors for autism: do they help cause de novo genetic mutations that contribute to the disorder? *Medical Hypotheses.* 2010;74(1):102–106; Kalkbrenner AE, Schmidt RJ, Penlesky AC. Environmental chemical exposures and autism spectrum disorders: a review of the epidemiological evidence. *Current Problems in Pediatric and Adolescent Health Care.* 2014;44(10):277–318.

35. Superfund: National Priorities List. US Environmental Protection Agency. www.epa.gov/superfund/superfund-national-priorities-list-npl. Updated on June 4, 2018. Accessed January 9, 2019.

36. Wertz J. EPA vows to speed cleanup of toxic Superfund sites despite funding drop [transcript]. National Public Radio. October 11, 2017. www.npr.org/2017/10/11/554564288/epa-vows-to-speed-cleanup-of-toxic-superfund-sites-despite-funding-drop. Accessed June 28, 2019.

37. Anderson B. Taxpayer dollars fund most of the oversight and cleanup costs at Superfund sites. *Washington Post.* September 20, 2017. www .washingtonpost.com/national/taxpayer-dollars-fund-most-oversight-and -cleanup-costs-at-superfund-sites/2017/09/20/aedcd426-8209-11e7-902a -2a9f2d808496_story.html?utm_term=.ed42f9287636. Accessed June 28, 2019.

38. Bienkowski B. Carcinogenic chemical spreads beneath American town. *Scientific American.* September 3, 2013. www.scientificamerican.com/article /carcinogenic-chemical-spreads-beneath-american-town. Accessed June 28, 2019.

39. Background on drinking water standards in the Safe Drinking Water Act (SDWA). US Environmental Protection Agency. www.epa.gov/dw standardsregulations/background-drinking-water-standards-safe-drinking -water-act-sdwa. Updated February 8, 2017. Accessed March 30, 2019.

40. Ibid.

41. Perlman H. Contamination in U.S. private wells. US Geological Survey. http://water.usgs.gov/edu/gw-well-contamination.html. Updated December 2, 2016. Accessed March 30, 2019.

42. Trichloroethylene (TCE) contamination in public water systems nationwide: 2015. Environmental Working Group. www.ewg.org/interactive -maps/2018-tce. Accessed March 30, 2019.

43. Protect your home's water. US Environmental Protection Agency. www .epa.gov/privatewells/protect-your-homes-water#welltestanchor. Updated December 15, 2017. Accessed March 29, 2019.

44. Sharp R, Pestano JP. Water treatment contaminants. Environmental Working Group. February 27, 2013. www.ewg.org/research/water-treatment -contaminants. Accessed March 30, 2019.

45. Choosing home water filters and other water treatment systems. Centers for Disease Control and Prevention. www.cdc.gov/healthywater/drinking /home-water-treatment/water-filters/step2.html. Updated June 3, 2014. Accessed March 30, 2019.

46. Ibid.

47. Patient groups applaud spending bill that funds important neuro data system. National Multiple Sclerosis Society. September 28, 2018. www .nationalmssociety.org/About-the-Society/News/Patient-Groups-Applaud -Spending-Bill-That-Funds-Im. Accessed January 4, 2019.

48. Farmers, ranchers, and other agricultural managers. US Bureau of Labor Statistics. Occupational Outlook Handbook. www.bls.gov/ooh/management

/farmers-ranchers-and-other-agricultural-managers.htm. Accessed March 30, 2019.

49. Damalas CA, Koutroubas SD. Farmers' exposure to pesticides: toxicity types and ways of prevention. *Toxics* 2016;4(1):1.

50. Ibid.

51. Ibid.

52. Pesticide information profile: paraquat. Extension Toxicology Network. September 1993. http://pmep.cce.cornell.edu/profiles/extoxnet/metiram -propoxur/paraquat-ext.html. Accessed March 25, 2019.

53. Furlong M, Tanner CM, Goldman SM, et al. Protective glove use and hygiene habits modify the associations of specific pesticides with Parkinson's disease. *Environment International*. 2015;75:144–150.

54. Ibid.; Damalas CA, Koutroubas SD. Farmers' exposure to pesticides: toxicity types and ways of prevention. *Toxics* 2016;4(1):1.

55. Pezzoli G, Cereda E. Exposure to pesticides or solvents and risk of Parkinson disease. *Neurology*. 2013;80(22):2035–2041; Goldman SM, Quinlan PJ, Ross GW, et al. Solvent exposures and Parkinson disease risk in twins. *Annals of Neurology*. 2012;71(6):776–784; Bale AS, Barone S, Scott CS, Cooper GS. A review of potential neurotoxic mechanisms among three chlorinated organic solvents. *Toxicology and Applied Pharmacology*. 2011;255(1):113–126.

56. Trichloroethylene. US Environmental Protection Agency. April 1992. www.epa.gov/sites/production/files/2016-09/documents/trichloroethylene .pdf. Updated January 2000. Published 1992. Accessed March 28, 2019; Risk management for trichloroethylene (TCE). US Environmental Protection Agency. www.epa.gov/assessing-and-managing-chemicals-under-tsca /risk-management-trichloroethylene-tce. Updated December 14, 2017. Accessed November 16, 2018.

57. Trichloroethylene toxicity: where is trichloroethylene found? Agency for Toxic Substances & Disease Registry. November 8, 2007. www.atsdr.cdc .gov/csem/csem.asp?csem=15&po=5. Accessed March 30, 2019.

58. Trichloroethylene. National Toxicology Program. November 2016. www.niehs.nih.gov/health/materials/tce_508.pdf; OSH answers fact sheets: trichloroethylene. Canadian Centre for Occupational Health and Safety. www.ccohs.ca/oshanswers/chemicals/chem_profiles/trichloroethylene.html. Updated January 3, 2017. Accessed March 30, 2019.

59. Racette BA, McGee-Minnich L, Moerlein SM, Mink JW, Videen TO, Perlmutter JS. Welding-related parkinsonism. *Neurology*. 2001;56(1):8;

Sadek AH, Rauch R, Schulz PE. Parkinsonism due to manganism in a welder. *International Journal of Toxicology*. 2003;22(5):393–401; Flynn MR, Susi P. Neurological risks associated with manganese exposure from welding operations—a literature review. *International Journal of Hygiene and Environmental Health*. 2009;212(5):459–469.

60. Welders, cutters, solderers, and brazers. US Bureau of Labor Statistics. Occupational Outlook Handbook. www.bls.gov/ooh/production/welders -cutters-solderers-and-brazers.htm. Accessed March 30, 2019.

61. Estruch R, Ros E, Salas-Salvadó J, et al. Primary prevention of cardiovascular disease with a Mediterranean diet. *New England Journal of Medicine*. 2013;368(14):1279–1290; Singh RB, Dubnov G, Niaz MA, et al. Effect of an indo-Mediterranean diet on progression of coronary artery disease in high risk patients (Indo-Mediterranean Diet Heart Study): a randomised single-blind trial. *The Lancet*. 2002;360(9344):1455–1461.

62. Ascherio A, Zhang SM, Hernan MA, et al. Prospective study of caffeine consumption and risk of Parkinson's disease in men and women. *Annals of Neurology*. 2001;50(1):56–63; Ross GW, Abbott RD, Petrovitch H, et al. Association of coffee and caffeine intake with the risk of Parkinson disease. *JAMA*. 2000;283(20):2674–2679.

63. Haspel T. The truth about organic produce and pesticides. *Daily Gazette*. May 23, 2018.

64. Blackwell DL, Clarke TC. State variation in meeting the 2008 federal guidelines for both aerobic and muscle-strengthening activities through leisure-time physical activity among adults aged 18–64: United States, 2010–2015. National Health Statistics Report. 2018;Jun(112):1–22.

65. Schenkman M, Moore CG, Kohrt WM, et al. Effect of high-intensity treadmill exercise on motor symptoms in patients with de novo Parkinson disease: a phase 2 randomized clinical trial. *JAMA Neurology*. 2018;75(2):219–226; Miller Koop M, Rosenfeldt AB, Alberts JL. Mobility improves after high intensity aerobic exercise in individuals with Parkinson's disease. *Journal of the Neurological Sciences*. 2019;399:187–193.

66. Snijders AH, van Kesteren M, Bloem BR. Cycling is less affected than walking in freezers of gait. *Journal of Neurology, Neurosurgery & Psychiatry*. 2012;83(5):575–576.

67. Hootman JM, Dick R, Agel J. Epidemiology of collegiate injuries for 15 sports: summary and recommendations for injury prevention initiatives. *Journal of Athletic Training*. 2007;42(2):311–319.

68. Gates RM. *Duty: Memoirs of a Secretary at War*. Rpt. ed: Vintage; 2015.

69. National Center for Injury Prevention and Control. *Report to Congress on Mild Traumatic Brain Injury in the United States: Steps to Prevent a Serious Public Health Problem.* Centers for Disease Control and Prevention; September 2003; National Center for Injury Prevention and Control. *Report to Congress on Traumatic Brain Injury in the United States: Epidemiology and Rehabilitation.* Centers for Disease Control and Prevention; 2015; The CDC, NIH, DoD, and VA Leadership Panel. *Report to Congress on Traumatic Brain Injury in the United States: Understanding the Public Health Problem among Current and Former Military Personnel.* Centers for Disease Control and Prevention (CDC), the National Institutes of Health (NIH), the Department of Defense (DoD), and the Department of Veterans Affairs (VA); 2013.

70. McCrea M. *Mild Traumatic Brain Injury and Postconcussion Syndrome: The New Evidence for Diagnosis and Treatment.* Oxford University Press; 2008; Mac Donald CL, Johnson AM, Cooper D, et al. Detection of blast-related traumatic brain injury in U.S. military personnel. *New England Journal of Medicine.* 2011;364(22):2091–2100; Levin HS, Diaz-Arrastia RR. Diagnosis, prognosis, and clinical management of mild traumatic brain injury. *Lancet Neurology.* 2015;14(5):506–517; Andrews PJD, Sinclair HL, Rodriguez A, et al. Hypothermia for intracranial hypertension after traumatic brain injury. *New England Journal of Medicine.* 2015;373(25):2403–2412.

71. Maas AIR, Menon DK, Adelson PD, et al. Traumatic brain injury: integrated approaches to improve prevention, clinical care, and research. *Lancet Neurology.* 2017;16(12):987–1048.

72. Mission and goals. National Institutes of Health. www.nih.gov/about-nih/what-we-do/mission-goals. Updated July 27, 2017. Accessed March 30, 2019.

73. Marar M, McIlvain NM, Fields SK, Comstock RD. Epidemiology of concussions among United States high school athletes in 20 sports. *American Journal of Sports Medicine.* 2012;40(4):747–755; Lincoln AE, Caswell SV, Almquist JL, Dunn RE, Norris JB, Hinton RY. Trends in concussion incidence in high school sports: a prospective 11-year study. *American Journal of Sports Medicine.* 2011;39(5):958–963; Hootman JM, Dick R, Agel J. Epidemiology of collegiate injuries for 15 sports: summary and recommendations for injury prevention initiatives. *Journal of Athletic Training.* 2007;42(2):311–319.

74. Moses H, Matheson DM, Cairns-Smith S, George BP, Palisch C, Dorsey ER. The anatomy of medical research: US and international comparisons. *JAMA.* 2015;313(2):174–189.

75. Accelerating Medicines Partnership (AMP): Parkinson's disease. US Department of Health & Human Services. www.nih.gov/research-training /accelerating-medicines-partnership-amp/parkinsons-disease. Updated October 26, 2018. Accessed February 8, 2019.

76. Alzheimer's and dementia: facts and figures. Alzheimer's Association. www.alz.org/alzheimers-dementia/facts-figures. Accessed February 8, 2019.

77. Pfizer ends research for new Alzheimer's, Parkinson's drugs. Reuters. January 7, 2018. www.reuters.com/article/us-pfizer-alzheimers/pfizer-ends -research-for-new-alzheimers-parkinsons-drugs-idUSKBN1EW0TN. Accessed January 4, 2019.

78. Dunn A. Pfizer, Bain Capital launch neuroscience company backed with $350M. BioPharma Dive. October 23, 2018. www.biopharmadive.com /news/pfizer-bain-capital-neuroscience-Cerevel/540359. Accessed February 8, 2019.

79. Moses H, Matheson DM, Cairns-Smith S, George BP, Palisch C, Dorsey ER. The anatomy of medical research: US and international comparisons. *JAMA*. 2015;313(2):174–189.

80. American Parkinson Disease Association. www.apdaparkinson.org /about-apda. Accessed March 30, 2019.

81. Brian Grant Foundation. https://briangrant.org. Accessed March 30, 2019.

82. *Parkinson's Foundation 2017 Annual Report*. Parkinson's Foundation; 2017.

83. Impact of the Ice Bucket Challenge. ALS Association. www.alsa.org /fight-als/ibc-infographic.html. Accessed February 8, 2019.

84. ACT UP New York. www.actupny.org. Accessed March 30, 2019.

85. ParkinsonNet. www.parkinsonnet.info. Accessed January 5, 2019.

86. Okun MS, Ramirez-Zamora A, Foote KD. Neuromedicine service and science hub model. *JAMA Neurology*. 2018;75(3):271–272.

87. Bloem BR, Rompen L, de Vries NM, Klink A, Munneke M, Jeurissen P. ParkinsonNet: a low-cost health care innovation with a systems approach from the Netherlands. *Health Affairs*. 2017;36:1987–1996.

88. van der Kolk NM, de Vries NM, Penko AL, et al. A remotely supervised home-based aerobic exercise programme is feasible for patients with Parkinson's disease: results of a small randomised feasibility trial. *Journal of Neurology, Neurosurgery & Psychiatry*. 2018;89(9):1003.

89. Szanton SL, Leff B, Wolff JL, Roberts L, Gitlin LN. Home-based care program reduces disability and promotes aging in place. *Health Affairs*. 2016;35(9):1558–1563.

90. Buchanan RJ, Wang S, Huang C, Simpson P, Manyam BV. Analyses of nursing home residents with Parkinson's disease using the minimum data set. *Parkinsonism & Related Disorders.* 2002;8(5):369–380; Ornstein KA, Leff B, Covinsky KE, et al. Epidemiology of the homebound population in the United States. *JAMA Internal Medicine.* 2015;175(7):1180–1186; Mitchell SL, Kiely DK, Kiel DP, Lipsitz LA. The epidemiology, clinical characteristics, and natural history of older nursing home residents with a diagnosis of Parkinson's disease. *Journal of the American Geriatrics Society.* 1996;44(4):394–399.

91. Centers for Medicare & Medicaid Services. Medicare & Home Health Care. US Department of Health and Human Services. October 2017. www .medicare.gov/sites/default/files/2018-07/10969-medicare-and-home -health-care.pdf; Your Medicare coverage. Medicare.gov. www.medicare.gov /coverage/home-health-services. Accessed March 30, 2019.

92. Woods B. By 2020, 90% of world's population aged over 6 will have a mobile phone: report. The Next Web. November 18, 2014. https:// thenextweb.com/insider/2014/11/18/2020-90-worlds-population-aged-6 -will-mobile-phone-report. Accessed October 11, 2018.

93. Dorsey ER, Glidden AM, Holloway MR, Birbeck GL, Schwamm LH. Teleneurology and mobile technologies: the future of neurological care. *Nature Reviews Neurology.* 2018;14:285–297.

94. WHO Model List of Essential Medicines. World Health Organization; April 2015.

95. Clinical trial delays: America's patient recruitment dilemma. Drug Development Technology. July 18, 2012. www.drugdevelopment-technology .com/features/featureclinical-trial-patient-recruitment. Accessed March 30, 2019.

96. Fox Trial Finder. The Michael J. Fox Foundation for Parkinson's Research. https://foxtrialfinder.michaeljfox.org. Accessed March 30, 2019.

97. Why Clinical Trials? The Michael J. Fox Foundation for Parkinson's Research. https://files.michaeljfox.org/WhyClinicalTrials.pdf. Accessed January 3, 2019; Schneider MG, Swearingen CJ, Shulman LM, Ye J, Baumgarten M, Tilley BC. Minority enrollment in Parkinson's disease clinical trials. *Parkinsonism & Related Disorders.* 2009;15(4):258–262.

98. Bot BM, Suver C, Neto EC, et al. The mPower study, Parkinson disease mobile data collected using ResearchKit. *Scientific Data.* 2016;3:160011.

99. Fox Insight. The Michael J. Fox Foundation for Parkinson's Research. https://foxinsight.michaeljfox.org. Accessed January 6, 2019.

101. Berg D, Postuma RB, Adler CH, et al. MDS research criteria for pro-dromal Parkinson's disease. *Movement Disorders.* 2015;30(12):1600–1611.

101. Pearson SD, Rawlins MD. Quality, innovation, and value for money: NICE and the British National Health Service. *JAMA.* 2005;294(20):2618–2622.

102. Weintraub A. How to cover Novartis' $475K CAR-T drug Kymriah? A "new payment model" is the only way, Express Scripts says. FiercePharma. September 22, 2017. www.fiercepharma.com/financials/car-t-and-other-gene -therapies-need-new-payment-model-says-express-scripts. Accessed January 6, 2019; Rosenblatt M, Termeer H. Reframing the conversation on drug pric-ing. *NEJM Catalyst.* November 20, 2017. https://catalyst.nejm.org/reframing -conversation-drug-pricing. Accessed June 28, 2019.

103. Danzon PM, Towse A. Differential pricing for pharmaceuticals: reconciling access, R&D and patents. *International Journal of Health Care Finance and Economics.* 2003;3(3):183–205; Hammer PJ. Differential pric-ing of essential AIDS drugs: markets, politics and public health. *Journal of International Economic Law.* 2002;5(4):883–912.

POSTSCRIPT

1. Pei S, Kandula S, Shaman J. Differential Effects of Intervention Timing on COVID-19 Spread in the United States. *medRxiv.* 2020.

2. Pesticide National Synthesis Project: Estimated annual agricultural pesticide use. *US Geological Survey.* www.water.usgs.gov/nawqa/pnsp/usage /maps/show_map.php?year=2017& map=PARAQUAT&hilo=H. Updated June 18, 2020, Accessed October 25, 2020.

3. Meyer B. Update from D.C.: EPA Allows Usage of Hazardous Pesti-cide Paraquat to Continue in the U.S. *Michael J. Fox Foundation.* October 26, 2020. www.michaeljfox.org/news/update-dc-epa-allows-usage-hazardous -pesticide-paraquat-continue-us. Accessed November 2, 2020.

4. Martin J. TOXMAP, federal database allowing public to track U.S. pollution, shut down after 15 years by National Library of Medicine. *News-week Digital.* December 24, 2019. www.newsweek.com/toxmap-federal -database-allowing-public-track-us-pollution-shut-down-after-15-years -trump-1479146. Accessed October 26, 2020.

5. Peters R, Ee N, Peters J, Booth A, Mudway I, Anstey KJ. Air Pollu-tion and Dementia: A Systematic Review. *Journal of Alzheimer's Disease.* 2019;70(s1):S145–S163.

6. Van Der Boom N. Doek valt definitief voor Mancozeb [Cloth falls definitively for Mancozeb]. *Boerenbusiness.* October 26, 2020. www.boeren

business.nl/akkerbouw/artikel/10889806/doek-valt-definitief-voor
-mancozeb. Accessed November 2, 2020.

7. GovTrack.us. S. 4406—116th Congress: Protect America's Children from Toxic Pesticides Act. 2020. www.govtrack.us/congress/bills/116
/s4406. Accessed October 25, 2020; U.S. Congress, House. *Protect America's Children from Toxic Pesticides Act.* HR 7940, 116th Congress. Introduced August 4, 2020. www.congress.gov/bill/116th-congress/house-bill/7940
/all-info?r=7&s=1. Accessed October 25, 2020; Udall, Neguse Introduce Landmark Pesticide Reform to Protect Children, Farmworkers and Consumers from Toxic Pesticides. *Tom Udall Senator for New Mexico.* www.tomudall
.senate.gov/news/press-releases/udall-neguse-introduce-landmark-pesticide
-reform-to-protect-children-farmworkers-and-consumers-from-toxic-pestici
des. Accessed October 25, 2020.

8. Marshal M. How COVID-19 can damage the brain. *Nature.* Online news feature. 2020;585(1476–4687):342–343. www.nature.com/articles
/d41586-020-02599-5. Accessed October 26, 2020.

9. Cohen ME, Eichel R, Steiner-Birmanns B, Janah A, Ioshpa M, Bar-Shalom R, Paul JJ, Gaber H, Skrahina V, Bornstein NM, Yahalom G. A case of probable Parkinson's disease after SARS-CoV-2 infection. *The Lancet Neurology.* 2020;19(10):804–805; Faber I, Brandão PRP, Menegatti F, De Carvalho Bispo DD, Maluf FB, Cardoso F. Coronavirus Disease 2019 and Parkinsonism: A Non-post-encephalitic Case. *Movement Disorders: Official Journal of the Movement Disorder Society.* 2020;35(10):1721–1722.

10. The federal response to COVID-19. *USAspending.gov.* www.usa
spending.gov/disaster/covid-19. Accessed October 25, 2020.

11. Estimates of funding for various research, condition, and disease categories (RCDC). *NIH Research Portfolio Online Reporting Tools.* February 24, 2020. www.report.nih.gov/categorical_spending.aspx. Accessed October 25. 2020.

12. Verma S. Early impact of CMS expansion of Medicare teleheath during COVID-19. *Health Affairs Blog.* July 15, 2020. www.healthaffairs.org
/do/10.1377/hblog20200715.454789/full/ Accessed October 25, 2020.

13. Cilia R, Mancini F, Bloem BR, Eleopra R. Telemedicine for parkinsonism: A two-step model based on the COVID-19 experience in Milan, Italy. *Parkinsonism & Related Disorders.* 2020;75:130–132; Bloem BR, Dorsey ER, Okun MS. The Coronavirus Disease 2019 Crisis as Catalyst for Telemedicine for Chronic Neurological Disorders. *JAMA Neurology.* 2020;77(8):927–928.

14. Lam K, Lu AD, Shi Y, Covinsky KE. Assessing Telemedicine Unreadiness Among Older Adults in the United States During the COVID-19 Pandemic. *JAMA Internal Medicine.* 2020;180(10):1389–1391.

15. Dahodwala N, Siderowf A, Xie M, Noll E, Stern M, Mandell DS. Racial differences in the diagnosis of Parkinson's disease. *Movement Disorders: Official Journal of the Movement Disorder Society.* 2020;24(8):1200–1205; Dahodwala N, Xie M, Noll E, Siderowf A, Mandell DS. Treatment disparities in Parkinson's disease. *Annals of Neurology,* 2009;66(2):142–145.

16. Hobbs TD, Hawkins L. The results are in for remote learning: it didn't work. *Wall Street Journal.* June 5, 2020. www.wsj.com/articles/schools -coronavirus-remote-learning-lockdown-tech-11591375078. Accessed October 25, 2020.

Index

Ray Dorsey is the David M. Levy Professor of Neurology at the University of Rochester, where he directs the Center for Health + Technology. For the past decade, he has used telemedicine to improve care for individuals with Parkinson's disease. His research has been published in the leading neurology, medical, and economic journals and been featured on NPR and in the *Wall Street Journal* and *New York Times*. Previously, he directed the Parkinson's disease division at Johns Hopkins Medicine and consulted for McKinsey & Company.

Todd Sherer is chief executive officer of The Michael J. Fox Foundation for Parkinson's Research, the largest and most highly respected private funder of Parkinson's disease research in the world. He is a neuroscientist who conducted pioneering work at Emory University that linked a common pesticide to Parkinson's disease. He earned his PhD at the University of Virginia and his BS at Duke University.

Michael S. Okun is the Adelaide Lackner Professor and Chair of Neurology at the University of Florida. He is one of the world's leading Parkinson's researchers and has advanced surgical treatments for the disease. He is the national medical director of the Parkinson's Foundation, the country's largest Parkinson's disease patient and advocacy group. He is a prolific writer with over five hundred publications and author of the book *Parkinson's Treatment: 10 Secrets to a Happier Life*, which has been translated into twenty languages and is a best seller for patients and families.

Bastiaan R. Bloem is professor of neurology and director of the Centre of Expertise for Parkinson & Movement Disorders at Radboud University Medical Centre in Nijmegen, the Netherlands. In 2004, with Dr. Marten Munneke, he created ParkinsonNet, the largest integrated-care program for Parkinson's patients. He is co–editor in chief of the *Journal of Parkinson's Disease* and secretary of the International Parkinson and Movement Disorder Society, has over six hundred publications, and has been a featured speaker at leading patient, academic, and health conferences globally.

PublicAffairs is a publishing house founded in 1997. It is a tribute to the standards, values, and flair of three persons who have served as mentors to countless reporters, writers, editors, and book people of all kinds, including me.

I. F. STONE, proprietor of *I. F. Stone's Weekly*, combined a commitment to the First Amendment with entrepreneurial zeal and reporting skill and became one of the great independent journalists in American history. At the age of eighty, Izzy published *The Trial of Socrates*, which was a national bestseller. He wrote the book after he taught himself ancient Greek.

BENJAMIN C. BRADLEE was for nearly thirty years the charismatic editorial leader of *The Washington Post*. It was Ben who gave the *Post* the range and courage to pursue such historic issues as Watergate. He supported his reporters with a tenacity that made them fearless and it is no accident that so many became authors of influential, best-selling books.

ROBERT L. BERNSTEIN, the chief executive of Random House for more than a quarter century, guided one of the nation's premier publishing houses. Bob was personally responsible for many books of political dissent and argument that challenged tyranny around the globe. He is also the founder and longtime chair of Human Rights Watch, one of the most respected human rights organizations in the world.

. . .

For fifty years, the banner of Public Affairs Press was carried by its owner Morris B. Schnapper, who published Gandhi, Nasser, Toynbee, Truman, and about 1,500 other authors. In 1983, Schnapper was described by *The Washington Post* as "a redoubtable gadfly." His legacy will endure in the books to come.

Peter Osnos, *Founder*